MIX
Papier aus verantwortungsvollen Quellen
Paper from responsible sources
FSC® C105338

Dr. Md. Iqbal Hossain (Editor)
Nadia Sultana

Simulation of Mass Transfer Phenomenon in a CAD Drug Eluting Stent System

Anchor Academic Publishing

Hossain, Md. Iqbal (Ed.), Sultana, Nadia: Simulation of Mass Transfer Phenomenon in
a CAD Drug Eluting Stent System, Hamburg, Anchor Academic Publishing 2017

Buch-ISBN: 978-3-96067-166-4
PDF-eBook-ISBN: 978-3-96067-666-9
Druck/Herstellung: Anchor Academic Publishing, Hamburg, 2017

Bibliografische Information der Deutschen Nationalbibliothek:
Die Deutsche Nationalbibliothek verzeichnet diese Publikation in der Deutschen
Nationalbibliografie; detaillierte bibliografische Daten sind im Internet über
http://dnb.d-nb.de abrufbar.

Bibliographical Information of the German National Library:
The German National Library lists this publication in the German National Bibliography.
Detailed bibliographic data can be found at: http://dnb.d-nb.de

All rights reserved. This publication may not be reproduced, stored in a retrieval system
or transmitted, in any form or by any means, electronic, mechanical, photocopying,
recording or otherwise, without the prior permission of the publishers.

Das Werk einschließlich aller seiner Teile ist urheberrechtlich geschützt. Jede Verwertung
außerhalb der Grenzen des Urheberrechtsgesetzes ist ohne Zustimmung des Verlages
unzulässig und strafbar. Dies gilt insbesondere für Vervielfältigungen, Übersetzungen,
Mikroverfilmungen und die Einspeicherung und Bearbeitung in elektronischen Systemen.

Die Wiedergabe von Gebrauchsnamen, Handelsnamen, Warenbezeichnungen usw. in
diesem Werk berechtigt auch ohne besondere Kennzeichnung nicht zu der Annahme,
dass solche Namen im Sinne der Warenzeichen- und Markenschutz-Gesetzgebung als frei
zu betrachten wären und daher von jedermann benutzt werden dürften.

Die Informationen in diesem Werk wurden mit Sorgfalt erarbeitet. Dennoch können
Fehler nicht vollständig ausgeschlossen werden und die Diplomica Verlag GmbH, die
Autoren oder Übersetzer übernehmen keine juristische Verantwortung oder irgendeine
Haftung für evtl. verbliebene fehlerhafte Angaben und deren Folgen.

Alle Rechte vorbehalten

© Anchor Academic Publishing, Imprint der Diplomica Verlag GmbH
Hermannstal 119k, 22119 Hamburg
http://www.diplomica-verlag.de, Hamburg 2017
Printed in Germany

AUTHOR'S ACKNOWLEDGMENTS

First of all I would like to pay my gratitude to Almighty Allah for Her loving care and guidance throughout our life.

Then, I would like to express my profound respect to my research supervisor Dr. Md. Iqbal Hossain, Assistant Professor, Department of Chemical Engineering, BUET for his valuable guidance, constant encouragement and keen interest at every stage of this study, without which it would have been extremely difficult to accomplish this research work.

I am very much grateful to Abbott Vascular, especially to Dr. Syed Hossainy for providing technical and financial aid to this work.

I would also like to thank the Department of Chemical Engineering, BUET for providing the necessary research facility.

I am grateful to Dr. Zakir Hossain, Assistant Professor, Department of Mechanical Engineering, BUET, for his vital inspiration and providing me the major theoretical concept and constructive suggestion on this work.

Last, but not the least, I want to thank my parents for their love, support, encouragement and prayer.

Nadia Sultana

CONTENTS

Chapter One	Introduction	1
1.1	Background of the Study	1
1.2	Objective of the Study	3
1.3	Scope of the Study	4
1.4	Thesis Organization	4
Chapter Two	**Literature Review**	**6**
2.1	Background of Current DES Technology	7
2.2	Current State of the Art: DES	11
	2.2.1 First Generation DES	11
	2.2.2 Second Generation DES	13
	2.2.3 Bioresorbable Polymer Stents	15
2.3	Long-term DES Safety	17
2.4	Other Uses of Stents	18
	2.4.1 Stents in Urology	18
	2.4.2 Stents for management of Tracheobronchial Obstruction	19
	2.4.3 Stents in the Esophagus and Gastrointestinal Tract	19
2.5	Theoretical Models to Describe Drug Releasing Behavior	19
2.6	Conclusion	20
Chapter Three	**Objective with Specific Aim and Research Significance**	**21**
3.1	Objective of Study with Specific Aims	21
3.2	Research Significance	22
Chapter Four	**Research Methodology**	**23**
4.1	Governing Equation of the Problem	23
	4.1.1 Dimensionless Parameter of Drug Release	28
	4.1.2 Diffusion in Porous Material	29
	4.1.3 Artery Wall Classification	30

	4.2	Computational Fluid Dynamics	31
		4.2.1 Finite Volume Method	32
	4.3	Simulation of the Governing Equation	33
		4.3.1 Solution of 1D Unsteady Concentration Equation	33
		4.3.2 Solution of 2D Unsteady Concentration Equation	39
		4.3.3 Solution of Concentration Equation of BSTM	47
		4.3.4 Solution of Equation Describing Coating Thickness	52
Chapter Five		**Results and Discussion**	**53**
	5.1	Solution of the Coating Thickness Equation	53
	5.2	Solution of Unsteady 1D Concentration Governing Equation	68
		5.2.1 Drug Concentration with respect to Radial Position	70
		5.2.2 Drug Concentration with respect to Time	75
	5.3	Solution of Unsteady 2D Concentration Governing Equation	79
		5.3.1 Drug Concentration Variation at Time-Radius Plane	80
		5.3.2 Drug Concentration Variation at r-z Plane	86
		5.3.3 Drug Concentration Variation at time-z Plane	94
	5.4	Grid Independency Test for the Models	98
		5.4.1 Grid Independency Test for 1D Concentration Model	98
		5.4.2 Grid Independency Test for 2D Concentration Model	99
Chapter Six		**Conclusion**	**101**
	6.1	Conclusions	101
	6.2	Recommendations for Future Work	102
References			**103**
Appendix A		Some Necessary Theories of Finite Volume Method	**116**

Chapter One

INTRODUCTION

1.1 Background of the Study

Coronary artery disease is the most common type of heart diseases and the leading cause of death worldwide due to heart disease. It occurs when the arteries that supply blood to the heart become narrowed or blocked by a buildup of cholesterol and other material called plaque at the inner wall of the artery as shown in Fig.1.1. This buildup is called atherosclerosis. With time, due to atherosclerosis less blood can flow through the arteries which results an insufficient supply of blood and oxygen to heart, initiating chest pain: angina. Limitation of blood flow to the heart causes ischemia (cell starvation secondary to a lack of oxygen) of the myocardial cells. Myocardial cells may die from lack of oxygen and this is called a myocardial infarction (commonly called a heart attack). It leads to heart muscle damage, heart muscle death and later myocardial scarring without heart muscle re-growth. Chronic high-grade restenosis of the coronary arteries can induce transient ischemia which leads to the induction of a ventricular arrhythmia, which may terminate into ventricular fibrillation leading to death. Myocardial infarction usually results from the sudden occlusion of a coronary artery when a plaque ruptures; activating the clotting system and atheroma-clot interaction fills the lumen of the artery to the point of sudden closure. The narrowing of the lumen of the heart artery before sudden closure is often not severe. The events leading up to plaque rupture are not understood despite many theories. Myocardial infarction is almost never caused by temporary spasm of the artery wall occluding the lumen, a condition also associated with atheromatous plaque and CAD.

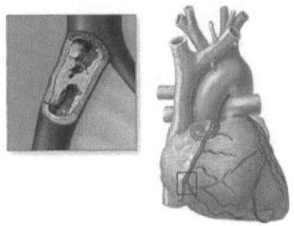

Fig. 1.1: Atherosclerosis [1]

When blockages in the coronary arteries develop, some symptoms like chest pain or pressure and/or shortness of breath are found whose causes have been mentioned above. Treatment for this condition (coronary artery disease) will depend on the type of the blockage and its extent. Treatment options include medication, surgery known as coronary artery bypass surgery, or catheter-based procedures. Several types of catheter-based procedures are available. During balloon angioplasty, a special balloon catheter is passed into the narrowed segment of the artery and expands the balloon, which thus opens the artery and compresses the blockage against the wall of the artery. More than one third of patients who undergo balloon angioplasty may experience restenosis or re-narrowing of the diseased artery segment within 6 months of the procedure. Stents are very small metal mesh-tubes that can be inserted via a balloon catheter into the narrowed segment of the artery as demonstrated in Fig.1.2. When the balloon is inflated, the stent expands and is embedded into the artery vessel wall, which thus opens the previously narrowed segment of artery. The balloon is then deflated and removed along with the catheter, and the stent is left behind to serve as a metal framework for the artery. Although stented arteries have less chance of re-narrowing than arteries opened with a balloon alone, in-stent restenosis can still occur in more than 1 in 5 patients after stent placement because, restenosis within the stented region of a heart artery is caused by tissue growth as well as inflammation. Thus some stents called drug-eluting stents have medication on them to inhibit or prevent this tissue growth. Drug-eluting stents are placed in a fashion similar to other stents; however, their use markedly reduces the rate of re-narrowing. In fact, about 1 in 10 patients develops re-narrowing in the several years after drug-eluting stent implantation, a rate about half of that seen for stents without medication.

As stents expose some foreign material to the blood stream, a small risk exists that a blood clot may develop in the stent, a process called stent thrombosis. These blood clots can occur many months and even years after stent implantation and may lead to a heart attack or death. All stents can potentially be affected by stent thrombosis. For this reason, most patients with stents are instructed to take anti-clotting medication, usually a combination of aspirin and clopidogrel or ticlopidine. Each of these medications stops platelet (particles in the blood that help clots to form) formation with functioning to their full capacity. The precise duration of anti-clotting medication dose depends on the type of stent placed.

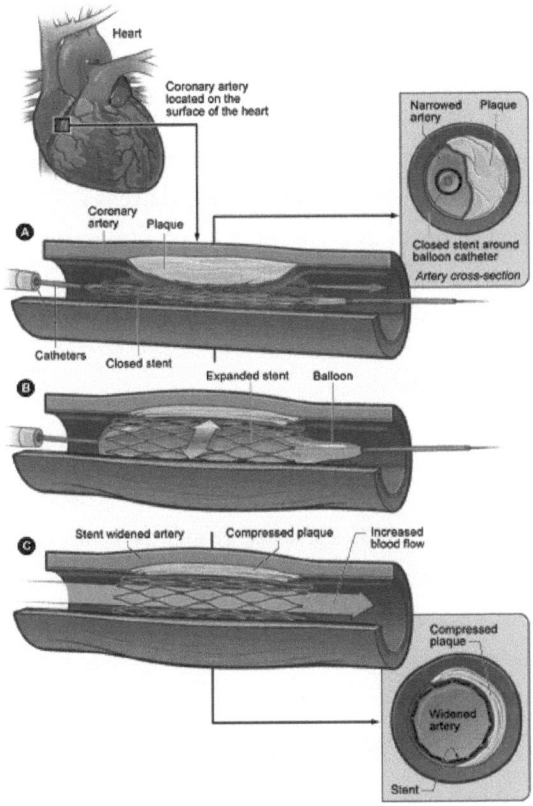

Fig. 1.2: Placing of stent in a diseased artery [2]

1.2 Objective of the study

In case of drug eluting stents a certain amount of anti-flammating drug is loaded in the coating over the base stent. This drug is released at the wall of diseased artery so that restenosis cannot take place at the place of artery where stent has been implanted. The releasing of the drug from the coating depends on the coating type given and drug loading, drug concentration in the coating and artery tissue properties. Though the drug given in the coating is only intended to release in the artery wall still some drug may release to the blood which may complexify the calculation of drug concentration developed in artery wall. In general, the mass transfer phenomenon occurs in a CAD-

DES system is still not fully understood and the studies available in literature is extremely limited.

Therefore, the main objective of this thesis was to simulate the overall mass transfer phenomenon occurs in CAD-DES system in the treatment of coronary artery disease. Where, decay of drug coating thickness and change in drug concentration in the artery tissue layer with time was evaluated. For this purpose the drug concentration in the artery tissue layer was described using a second order partial differential equation which was then solved using finite volume algorithm of computational fluid dynamics.

1.3 Scope of the Study

In this thesis work a drug eluting stent was studied where there was a biodegradable coating over a bare metal stent in which there was some amount of therapeutic drug. The degradation of the biodegradable coating layer thickness was determined with respect to time which was actually representing the remaining drug concentration in the coating layer. Then using this variable drug concentration as the drug concentration at initial tissue layer concentration profile of drug in tissue layer with respect to time and position was determined using finite volume algorithm, where this algorithm was coded using MATLAB programming language.

1.4 Thesis Organization

This thesis contains six chapters. Those are organized as follows:

Chapter One contains introduction of the thesis. Summary of the background, objective and scope of the study and thesis organization are included in the chapter.

Chapter Two provides an overview of the CAD and its treatment using DES. Several existing DES; background of this technology; safety, uses and future of DES are briefly are discussed here.

Chapter Three describes significance of DES in CAD treatment and specific aim of this thesis work.

Chapter Four details the research methodology of the work where, numerical background, governing equation, dependence of different parameters of the equation and elaborated solution technique has been included.

Chapter Five gives through idea of the research finding where all the graphs and data evaluated from the works has been included.

Chapter Six states the conclusion drawn from the current work and suggests possible directions for the future work.

Chapter Two

LITERATURE REVIEW

At present, coronary artery disease (CAD) is one of the leading causes of death and disability in the developed world. According to the American Heart Association CAD was responsible for approximately 445,687 deaths in the United States in 2005, representing 20% of all deaths that year [1]. CAD is caused by the development of atherosclerotic lesions within one or more of the coronary arteries which deliver oxygen and vital nutrients to the heart muscle. Several risk factors have been identified that contribute to the progression of this disease and include smoking, hypertension, diabetes and increased levels of cholesterol [1]. If the lumen becomes sufficiently narrowed, blood flow to a portion of the heart is restricted, usually resulting in angina pectoris. If untreated, vulnerable atherosclerotic lesions can become unstable and rupture. This often results in coronary occlusion and subsequent myocardial infarction.

Over the past two decades, percutaneous trans-luminal coronary angioplasty (PTCA) with bare-metal stent (BMS) placement has been utilized as a minimally invasive treatment for obstructive CAD. Typically, a BMS is a small, tubular, wire-mesh device which is pre-loaded in a collapsed form onto a catheter balloon, threaded to the narrowed section of the artery and expanded within the vessel. Once expanded, the BMS acts as a mechanical scaffold, reducing elastic recoil and maintaining vessel patency post-treatment. For many patients who suffer from CAD, treatment with a BMS will generally result in extremely favorable initial clinical results. However, at follow-up (6–12months), re-narrowing of the treated artery is commonly observed in 20–30% of patients [2]. This re-narrowing of the treated artery is due to in-stent restenosis (ISR) which is defined as diameter stenosis of ≥50% in the stented area of the vessel [3].

In recent years, DESs have been developed to address the problem of ISR. A DES typically consists of a BMS platform which has-been coated in a formulation of drugs and carrier materials. The drugs commonly employed are known to interrupt the key cellular and molecular processes associated with ISR. To date, clinical evaluation has overwhelmingly proven the superiority of DESs for the reduction of ISR rates compared to BMSs, leading to the regulatory approval of a number of DESs by both the European Union (EU) Conformities' European (CE) and the US Food and Drug Administration

(FDA). Despite the success of DESs in the treatment of CAD, concern has arisen over the long-term safety and efficacy of these devices due to cases of late adverse clinical events such as stent thrombosis. With this concern in mind, research and development in DES design is currently centered on increasing their performance and long-term safety. Though only five distinct DES have received both CE and FDA approval for commercial sale in theEU and the US, the number of DESs currently undergoing evaluation is substantial.

In Section 2.1, a background on current DES technology is provided. In Section 2.2, the current CE and FDA approved DESs are discussed in terms of their important design features and the clinical trials which led to their approval. In Section 2.3, long term safety of DES is discussed, in section 2.4 some other uses of DES has been given. At last in section 2.5 some model describing drug releasing behavior has been discussed.

2.1 Background of Current DES Technology

A stent is a medical device to serve as a temporary or permanent internal scaffold to maintain or increase the lumen of a body conduit. Metallic coronary stents were first introduced to prevent arterial dissections and to eliminate vessel recoil and intimal hyperplasia associated with percutaneous trans-luminal coronary angioplasty. It has now become the established mode of treatment of this type coronary intervention. It has been shown to reduce late restenosis relative to conventional balloon angioplasty [2] [5] [6].

Today, most stents employed by DESs are manufactured in modular or slotted-tube configurations and are delivered by balloon-dilation. The stent is crimped to a low-profile upon a balloon-tipped catheter and introduced to the cardiovascular system via the femoral or radial arteries. The stent must therefore have a low crimped profile and must possess a high level of flexibility to enable delivery through the tortuous cardiovascular system. During expansion the stent should experience minimum shortening and upon deployment should conform to the vessel geometry without straightening the vessel unnaturally. The stent should provide optimum vessel coverage and should possess high radial strength such that it undergoes minimal radial recoil and achieves a final diameter consistent with that of the host vessel upon unloading [7]. As the stent acts as a conduit for drug-delivery it is also important that its geometrical configuration facilitates homogeneous distribution of the drug within the vessel [8].

The metals used to prepare the stent are selected for strength, electricity and malleability or shape memory. Stainless steel, tantalum and nitinol alloys are among the most commonly used materials [9] [10] [11] [12]. Nitinol offers super-elastic and thermal shape memory properties which allow self-expansion of the stent during deployment and thermally induced collapse for theoretical removal procedure [13]. In recent years however, driven by emerging correlations between strut thickness and rates of ISR,metallic alloys such as cobalt–chromium have superseded steel as the material of choice for stent design [14]. These metallic alloys have been developed with increased levels of strength and X-ray attenuation compared to stainless steel, allowing newer stents to be designed with significantly thinner struts which do not impair the resulting strength, corrosion resistance or radiopacity of the device. Further development in stent design is currently centered on the assessment of stronger metallic alloys, compound metals andbioabsorbable materials.

The incidence of restenosis remains high despite technical and mechanical improvements. This restenosis is a result of in-stent neointiaml hyperplasia caused by proliferation and migration of vascular smooth muscle cells (VSMCs) induced by vessel wall injury [15]. The pathology of restenosis stems from a complex interaction between cellular and acellular elements of the vessel wall and the blood [16]. Some antiproliferative and anti-inflammatory agents have been shown to elute slowly from polymer coatings and to be associated with reduced neointimal formation in animal models.

Equally important as the actual drug or therapeutic agent that is released by aDESisthemechanismby which the drug is released. To date, the most successful method of facilitating drug adhesion and delivery from a stent has involved the use of permanent synthetic polymer coating materials such as polyethylene-co-vinyl acetate (PEVA), poly-n-butyl methacrylate (PBMA), and the tri-block copolymerpoly (styrene-b-isobutylene-b-styrene) (SIBS). By carefullymixing anti-restenotic drugs with these materials, a drug-polymer matrix may be formed and applied to the surface of the stent platform. Upon deployment, drug-delivery is driven by diffusion from the matrix and the rate of this diffusion is dictated by the type, composition and number of polymers used in the drug–polymer matrix.

In recent years these permanent polymers have been superseded by advanced biocompatible permanent polymers such asphosphorylchlorine (PC) and the co-polymerpoly (vinylidenefluoride-co-hexafluoropropylene) (PVDF-HFP). These advanced polymers mimic the phospholipids on the outer surfaces of red blood cells resulting in a stent platform that induces minimal thrombus formation upon deployment and has minimal adverse clinical effect on late healing of the vessel wall. Further development in this area is currently centered on the assessment of biocompatible and bioabsorbable polymer coating materials and on the development of novel mechanisms of drug release.

During the deployment of a DES, any mechanical injury incurred in the vessel leads to an immediate healing response in the arterial wall. This healing response is initially characterized by the activation of platelets within the intima, leading to thrombus formation and the recruitment of blood-borne monocytes, neutrophils and lymphocytes. These cells produce mitogenic and chemotactic factors which trigger the activation of smooth muscle cells (SMCs) which undergo unrestrained proliferation and migration toward the intimal layer resulting in neointimal growth and ISR [17]. As such, the ideal anti-restenotic agent should exhibit potent antiproliferativeeffects but preserve vascular healing. To date a vast number of immunosuppressive and anti-proliferative agents have been investigated for the prevention of ISR, however, only a small number have shown real effectiveness in clinical evaluation.

Two anti-proliferative agents, paclitaxel [18] [19] and sirolimus [20], have been used in humans with promising preliminary results. Paclitaxel is a natural or semi-synthetic diterpane composed of a rigid texane ring and a flexible side chain. It is an anti-neoplastic agent [21] [22]. However it is potentially cardiotoxic and the dose of paclitaxel that can be delivered safely has yet to be resolved [23] [24]. Paclitaxel has been shown to markedly attenuate stent-induced intimal thickening of the lumen [25] [26]. Paclitaxels'santiproliferative effect is reversible [27]. Its short cellular residence time: 1 hr, along with the reversible antiproliferative activity, suggests that it should be formulated in the sustained-release dosage form [28]. Sirolimus is a carbonyl lactone-lactam macrolide that has been shown to inhibit VSMC growth. This inhibition has been reported to be concentration dependent, with a threshold limit of 16.7 ng/ml [29].Sirolimus has been shown to be effective with a remarkable restenosis rate of almost 0% [30] [31] [32]. However, some criticism has been expressed regarding the absence of data in complex lesions, as well as long-term data [33]. Results of the RAVEL and SIRIUS trials demonstrated that sirolimus-eluting stents effectively inhibit restenosis in humans [34] [35]. The TAXUS trials revealed significant inhibition of coronary restenosis by paclitaxel [36] [37] [38] [39]. Drug eluting polymer coated stents have thus moved into the lime light as vehicles for the local drug administration [40].

In brief, Sirolimus, zotarolimus and everolimus, potent immunosuppressive agents, inhibit SMC proliferation in response to cytokine and growth factor stimulation by binding to the cytosolic FK binding protein 12 (FKBP12). This prevents the activation of the mammalian target of rapamycin (mTOR) and leads to interruption of the cell-cycle in the G1-S phase. Paclitaxel, a strong antiproliferative agent, suppresses neointimal growth by binding with and stabilizing microtubules. The stability of these microtubules inhibits their disassembly and renders them non-functional, resulting in cell-cycle arrest in the G0–G1 and G2–M phases (Fig. 2.1) [17]. Development in this area is currently centered on the assessment of further immunosuppressive and anti-proliferative agents as well as the evaluation of numerous migration-inhibiting, enhancedhealingand re-endothelialisation agents.

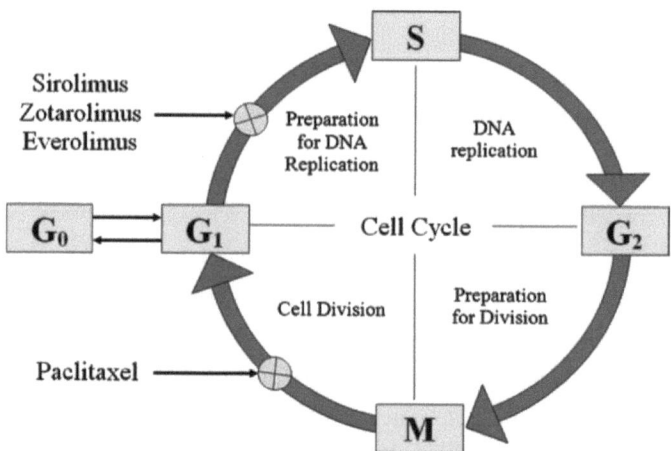

Fig. 2.1: Cell-cycle and mechanism of action of sirolimus, zotarolimus, everolimus and paclitaxel [4]

2.2 Current State of the Art: DES

Since 2002, five distinct DESs have received regulatory approval from the both the EU, CE and the USFDA: the first-generation Cypher sirolimus-eluting stent (SES) (Cordis, Johnson & Johnson, NJ, US), the Taxus Express2 paclitaxel-eluting stent (PES) (Boston Scientific, MS, US) and the TaxusLiberté PES (Boston Scientific), and the second generation Endeavor zotarolimus-eluting stent (ZES) (Medtronic Vascular, CA, US) and Xience-V everolimus-eluting stent (EES) (Abbott Vascular, CA, US).

2.2.1 First Generation DES

The Cypher SES consists of a Bx-Velocity BMS (Johnson & Johnson) coated in a formulation of sirolimus and two permanent polymers, PEVA and PBMA. The Bx-Velocity BMS is a closed-cell; slotted-tube stent manufactured from 316L stainless steel and is comprised of a series of sinusoidal strut-segments joined bin-shaped, flexible link-segments. The drug-polymer coating is applied to the entire stent surface with a standard concentration of 140μg of sirolimus per cm^2 of stent surface area and is designed to release approximately 80% of the drug within 30 days of stent deployment[41]. The Cypher SES is currently available in six lengths from 8 to 33mmand four diameters from 2.25 to 3.5mm. The principal safety and efficacy evidence for the Cypher SES was obtained from five clinical trials: the First In Man (FIM) trial, the RAVEL trial and the SIRIUS trials (SIRIUS, E-SIRIUS and C-SIRIUS).The FIM trial was a non-randomized trial involving 45 patients that demonstrated minimal in-stent neointimal proliferation with both fast- and slow-release SESs at 4 month follow-up [42]. The RAVEL trial was a randomized trial involving 238 patients with relatively low-risk lesions that demonstrated the superiority of theCypher SES over the Bx-Velocity BMS in terms of in-segment late loss at 6months [43]. The SIRIUS, C-SIRIUS and E-SIRIUS trials were randomized trials involving a total of 1510 patients with more complex lesions than those enrolled in the RAVEL and FIM trials. The superiority of the Cypher SES over the Bx-Velocity BMSwas furtherdemonstrated in these trials, with markedly lower rates of target lesion revascularization and adverse clinical events observed in patients treated with the Cypher SES [41][44][45]. The Cypher SESbecame the first DES to receive both CE and FDA approval in April2002 and April 2003, respectively.

The Taxus Express PES consists of an Express BMS (Boston Scientific) coated in a formulation of paclitaxel and a permanent co-polymer, SIBS. The Express BMS is a closed-cell; slotted-tube stent manufactured from 316L stainless steel and is comprised of a series of sinusoidal strut-segments joined by straight articulations to short, narrow strut-segments. The drug-polymer coating is applied to the entire stent surface in single layer with a standard concentration of 100µg of paclitaxel per cm^2 of stent surface area. The release of paclitaxel is bi-phasic with an early 48 h burst followed by a low-level release over the following 10 days [39]. TheTaxus Express2 PES is currently available in six lengths from 8 to 33mmand four diameters from 2.5 to 3.5mm. The principal safety and efficacy evidence for the Taxus Express2 PES was obtained from three clinical trials, the TAXUS I, II and IV trials. The TAXUS I trial was a randomized trial involving 61 patients which demonstrated zero binary in-stent restenosis with PESs at 6months and minimal adverse clinical events compared to BMSs at12 months [36]. The TAXUS II trial was a randomized trial involving536 patients that demonstrated the superiority of both slow- and fast-release PESs over BMSs in terms of stent volume obstructed byneointimal proliferation at 6 months [48]. The TAXUS IV trial was a randomized trial involving 1314 patients with more complex lesions than those enrolled in the TAXUS I and II trials that demonstrated the superiority of the Taxus Express2 PES over the Express BMS in terms of in-stent late loss, binary in-stent restenosisand target-lesion revascularization at 9 months [39]. The TaxusExpress2 PES became the second DES to receive both CE and FDAapproval in May 2002 and March 2004, respectively. Following FDA approval of the Taxus Express2 PES, the TAXUSclinical trial program was succeeded by the TAXUS ATLAS trial, designed to assess the drug-polymer coating (paclitaxel-SIBS)employed by the Taxus Express2 PES upon a new BMS platform,the Liberté stent (Boston Scientific). The Liberté stent is a closed-cell, slotted-tube stent manufactured from 316L stainless steel which has substantially thinner struts compared to the Express stent (0.097 vs. 0.132mm) allowing for improved flexibility and deliverability. The Liberté stent platform has also been specifically designed with a dense strut configuration which ensures homogeneous distribution of paclitaxel within the vessel. TheTaxusLiberté PES is currently available in seven lengths from 8 to38mm and five diameters from 2.25 to 4mm. The principal safety and efficacy evidence for the TaxusLiberté PES was obtained from the TAXUS ATLAS trial. The TAXUS ATLAS trial was a randomized trial involving 871patients that compared the safety and efficacy of the TaxusLibertéPES with an historic control arm of patients who were treated with a Taxus Express2 PES in the TAXUS IV and V trials. Despite a significantly higher incidence of complex lesions in the TAXUSATLAS patient population the TaxusLiberté PES was found to benon-inferior to the Taxus Express2 PES with similar rates of adverse clinical events, in-stent late loss and target-vessel revascularization observed at 9 months [49]. The TaxusLiberté PES received CE andFDA approval in September 2005 and October 2008, respectively.

2.2.2 Second Generation DES

The Endeavor ZES consists of a Driver BMS (Medtronic Vascular) coated in a formulation of zotarolimus and a biocompatible, permanent PC co-polymer. The Driver BMS is an open-cell; modular stent manufactured from MP35N cobalt–chromium and is comprised of a series of alternating upper and lower crowns connected by axial struts in a sinusoidal pattern. The use of MP35N cobalt–chromium alloy allows for relatively thin struts (0.091mm) to be used compared with first-generation DESs. The drug-polymer coating is applied to the entire stent surface with a standard concentration of 100μg of zotarolimus per cm of stent length and is designed to release approximately 95% of the total dose of zotarolimus within15 days of stent placement [50]. The Endeavor ZES is currently available in eight lengths from 8 to 30mm and three diameters from2.5 to 3.5mm. The principal safety and efficacy evidence for the Endeavor ZES was obtained from four clinical trials, the ENDEAVORI–IV trials. The ENDEAVOR I trial was a non-randomized trial involving 100patients that demonstrated the safety and efficacy of the EndeavorZES with minimal binary in-stent restenosis observed at four and12 months [50]. The ENDEAVOR II trial was a randomizedtrial involving 1197 patients that demonstrated the superiority of the Endeavor ZES over the Driver BMS with significantly lower rates of binary in-stent restenosis and target-vessel revascularization observed at 8 and 9 months, respectively [51]. The ENDEAVOR III (n = 436) and ENDEAVOR IV (n = 1548) trials were randomizedtrials designed to demonstrate the non-inferiority of the Endeavor ZES with the Cypher SES and Taxus Express2 PES, respectively. The initial performance of the Endeavor ZES in these trials was disappointing however, with markedly higher rates of target-lesion revascularization and significantly higher rates of in-stent late loss observed with the Endeavor ZES at short-term follow-up in both trials [52][53].Results at longer-term follow-up have been more reassuring however, with the absolute difference in target-lesion revascularization reduced to 1.6% at five years in the ENDEAVOR IIItrial and

0.5% at three years in the ENDEAVOR IV trial [54][55].The Endeavor ZES received both CE and FDA approval in July 2005 and February 2008, respectively.

The Xience-V EES consists of a Multi-Link Vision BMS (Abbott Vascular) coated in a formulation of everolimus, PBMA and a permanent biocompatible co-polymer, PVDF-HFP. The Multi-Link Vision BMS is a closed-cell, slotted-tube stent manufactured from L605 cobalt–chromium alloy and consists of a series of corrugated, zigzag strut-segments joined by single-turn link-segments. The use of L605 cobalt–chromium alloy allows for relatively thin struts (0.081mm) to be used compared with first-generation DESs. The drug-polymer coating is applied to the entire stent surface with a standard concentration of 100µg of everolimus per cm^2 of stent surface area and is designed to release approximately 80% of the total dose within 30 days of stent placement [56]. The Xience-VEES is currently available in six lengths from 8 to 28mm and five diameters from 2.5 to 4mm. The principal safety and efficacy evidence for the Xience-V EES was obtained from four clinical trials: the SPIRIT FIRST trial and the SPIRIT II–IV trials. The SPIRIT FIRST trial was a randomized trial involving 60patients that demonstrated the superiority of the Xience-V EESover the Multi-Link BMS in terms of in-stent late loss and binary in-stent restenosis at 6 months [56]. The SPIRIT II trial was a randomized trial involving 300 patients that demonstrated the superiority of the Xience-V EES over the Taxus Express2 PES in terms of in-stent late loss at 6 months [57]. The SPIRIT III trial was arandomized trial involving 1002 patients that demonstrated significantly reduced in-segment late loss and non-inferior rates of target-vessel failure in patients treated with a Xience-V EES compared to the Taxus Express2 PES at 12 months [58]. The SPIRIT Vitriol is a randomized trial involving 3687 patients that has demonstrated the superiority of the Xience-V EES over the Taxus Express2PES in terms of target-lesion failure and target-vessel revascularization at 12 months [59]. Interestingly, following three year follow-up of the SPIRIT II and III trials, investigators observed an increase in the absolute difference in target-vessel failure and adverse clinical events in favor of the Xience-V EES [60] [61]. The Xience-VEES received CE and FDA approval in January 2006 and July 2008, respectively.

Since obtaining CE and FDA approval, both first- and second generation DESs have been evaluated in dozens of clinical studies to assess their safety and efficacy when deployed in a number of patient and lesion sub-groups. These studies include evaluations of DES delivery in small vessels, long lesions, diabetics, chronic total occlusions (CTOs), bifurcated vessels, sapheneous vein grafts (SVGs), patients suffering from ISR, ST-elevated myocardial ischemia, multi-vessel disease and by direct delivery. These DESs have also been subject to a number of physician-driven registries to assess their relative and real-world performance [62].

2.2.3 Bioresorbable Polymer Stents

Bioresorbablepolymeric stents have attracted much attention as alternative to metallic stents. There are several reasons for fabricating a stent composed of a biodegradable polymeric material.Bioresorbable polymeric vascular stents have the potential to remain in situ for a predicted period of time, keeping the vessel wall patent and degrading to non-toxic substances. Accumulating evidence indicates that the use of a bioresorbable coronary stent dramatically decrease the need for a prosthesis after six months [63]. Bioresorbable stents are preferable for treatment of tracheomalacia in newborns and infants because removal surgeries are not necessary. Furthermore it can be used as support devices as well as platform for drug protein delivery to the conduit wall in all of the above mentioned applications.

PLLA, PGA, poly ε-caprolactone (PCL) and poly-D, L-lactic acid (PDLLA) are the most frequently used aliphatic poly (α-hydroxy-acids) for preparing bioresorbable stents [64] [65] [66]. The semi-crystalline PLLA and PGA have high initial tensile strength, permitting a robust mechanical design. The PLLA total degradation time is approximately 24 months, whereas that of PGA is 6-12 months. PLLA is one of the most important biodegradable polymers and is used in a wide range of clinical applications, including devices for orthopedic [67] [68] [69] and cardiovascular surgery [70], sutures [71] and drug delivering implants [72]. This polymer is very good choice, especially for the first three above mentioned applications, where high mechanical strength and toughness are required. PLLA can be formed into fibers, films, tubes and matrixes using standard processing techniques such as modeling, extrusion, spinning and solvent casting [71]. PCL is also a semi-crystalline polymer with a relatively high degree of crystallinity. However, it exhibits lower strength and modulus than PLLA and PGA owing to its low glass transition temperature. PDLLA is actually a random copolymer that consists of L-lactic acid and D-lactic acid monomers. It is therefore amorphous and cannot exhibit crystalline structures. Its strength and modulus are lower than those of PGA and PLLA. Polydioxanone (PDS) has gained increasing interest in the medical and pharmaceutical fields owing to its excellent biocompatibility [71]. Although it is semi-crystalline polymer it also exhibits lower strength than PLLA and PGA because it has low glass transition temperature, similar to PCL.

Table 2.1　Characteristics of typical bioresorbablepolymers[46]

Polymer	Melting point (^{0}C)	Glass transition Temperature (^{0}C)	Modulus (GPa)	Degradation time (months)
PGA	225 – 230	35 – 40	70	6 – 12
PLLA	173 – 178	60 – 65	2.7	> 24
PDLLA	Amorphous	55 – 60	1.9	12 – 16
PCL	58 – 63	(-65) – (-60)	0.4	> 24
PDS	N/A	(-10) – 0	1.5	6 – 12
85/15 PDLGA	Amorphous	55 – 55	2.0	5 – 6
75/25 PDLGA	Amorphous	50 – 55	2.0	4 – 5
50/50 PDLGA	Amorphous	45 – 50	2.0	1 – 2

Useful combination of the above mentioned materials can be created to alter the mechanical properties and drug release profiles of bioactive agents from polymeric structures based on these polymers. For example, 10/90 PDLGA, a random copolymer that contains 90% glycolic acid and 10% lactic acid has relatively high strength but is more flexible than PGA and degrades faster. Other PDLGA formulations that contain relatively high lactic acid contents, such as 85/15 PDLGA, 75/25 PDLGA and 50/50 PDLGA are amorphous therefore they do not exhibit high strength and modulus. However they can be used for support in case lower strength necessity such as neural stents.

All above mentioned materials degrade principally by simple hydrolysis of the ester bond in the polymer backbone. Partial chain scission degrades the polymer to 10 - 40μm particles. These particles can be phagocytized and metabolized to carbon dioxide and water, which are of course fully resorbed. The polymers degradation time is a function of its chemical structure and molecular weight. Crystallinity contributes to a higher degradation time because crystalline domains are denser than amorphous domains and water molecule cannot penetrate them easily. Table 2.1 demonstrates that a relatively small number of polymers provide a large variety of degradation times for various medical support applications.

2.3 Long-term DES Safety

Following the strong performance of the first-generation DESs in clinical evaluation, these devices were widely adopted by interventional cardiologists with up to 90% of stent procedures carried out in the US involving DES placement by late 2005 [74]. Over the following two years, however, major concerns arose over the long-term safety of DESs when a number of clinical and observational studies reported significantly increased risk of mortality in patients treated with DESs compared to BMSs beyond 12 months[75] [76] [77] [78]. Prompted by these results, a number of large-scale, meta-analyses were undertaken to assess both the short- and long-term safety of DESs relative to BMSs [79] [80] [81]. Reassuringly, no increased risk of mortality was observed between patients treated with DES and BMS with similar rates of death and myocardial infarction reported for DESs in each of these studies. Furthermore, in a resentment-analysis of long-term follow-up (1–4 years) from over 22 randomized clinical trials and 34 observational studies, patients treated with DESs were associated with lower rates of death and myocardial infarction and repeat revascularizations compared to patients treated with BMSs [82].

Today, the primary concern with long-term DES safety is stent thrombosis, a potentially fatal adverse event that often leads to myocardial infarction and/or death. In the debate that followed the initial concerns over the long-term safety of DESs it emerged that restrictive and non-uniform definitions of stent thrombosis had been utilized during the initial clinical evaluation of the firstgenerationDESs. The Academic Research Consortium subsequently recommendedstandardized definitions of stent thrombosis and in2007 these definitions were adopted in a pooled analysis of the long-term follow-up from eight clinical trials involving both the Cypher SES and the Taxus Express2 PES. Though similar rates of early (less than 1 month) and late (1–12 months) stent thrombosis were observed between DESs and BMSs in this analysis, higher rates of very-late (greater than 12 months) stent thrombosis were reported with DESs [83]. Evaluation of the long-term follow-up of clinical trials and registries has since supported this observation [84] [85] [86].

Although the exact cause of stent thrombosis is not yet fully understood a number of patient, lesion, and procedural factors have been associated with an increased risk of stent thrombosis. These include, increasing age, diabetes mellitus, renal failure, increasing stent and/or lesion length, decreasing stent and/or vessel diameter, treatment of bifurcation, treatment of CTO, treatment of ISR,stent under-expansion and premature discontinuation of dual anti-platelet therapy [87] [88]. Of note, recent studies have identified delayed healing and incomplete endothelial strut coverage as a primary risk factor for stent thrombosis [89] [90] [91] [92]. It has been shown that the non-erodible polymer coatings employed by DESs (particularly the first-generation Cypher SES and Taxus Express2 PES)impair stent strut endothelialisation and may induce late hypersensitivity reactions and subsequent stent thrombosis[93].As a result of these findings, research in this area is currently centered on the development and evaluation of improved DESs which maintain the impressive clinical benefits observed with currently approved devices while eradicating long-term safety concerns such as stent thrombosis.

2.4 Other uses of Stents

The range of stent application has expanded as more experience has been gained and following encouraging results in the treatment of vascular diseases. Stents have been used for treating urethral obstruction caused by benign prostatic hyperplasia and for treating benign or malignant tracheobronchial obstructions. They have also been use for supporting the neonatal trachea in tracheal malacia for treating benign and malignant esophageal, gastrointestinal and bile duct strictures and for treating arterial dissections, aneurysms and various neurovascular diseases.

2.4.1 Stents in Urology

Stents have been used to prevent urine retention following thermal treatment of benign prostatic hyperplasia (BPH) by various means, including trans-urethral microwave therapy and direct vision laser ablation of the prostate. Several stent designs were shown to prevent obstruction of the prostatic urethra and restructure of the anterior urethra. These stents include the Barnes, Finnish biodegradable self-reinforced polyglycolic acid (SR-PGA) spiral; the Nissenkorn; and the Trestle stents [94] [95] [96]. In clinical studies, researchers have used biodegradable stents to treat benign prostatic hyperplasia. The results obtained yield more positive outcomes compared with those using suprapublic catheters [97]. Self-reinforced poly (L-lactic acid) PLLA bioresorbable spiral stents are also undergoing evaluation for use in the anterior and posterior urethra and upper urinary tract for preventing urinary retention and for repairing local urethral trauma or defects [98] [99].

2.4.2 Stents for Managements of Tracheobronchial Obstruction

Techeobronchial obstruction owing to either benign or malignant disease causes significant morbidity and mortality. Metal stents which are originally developed for vascular system have been adapted for lesions involving the tracheobronchial tree and include the Gainturco-Z, Palmaz, Strecker, Ultraflex and Wallstent stents [100]. These stents were used successfully for treating patients with bronchogenic cancer, inoperable esophageal tumors, primary tracheal tumors and meta-static malignancies. Bioresorbable external tracheal stents have been investigated for treating pediatric tracheal malacia, for solving the problem of limited tracheal growth in children with rigid external fixation and for avoiding the necessity of a second procedure for removing the synthetic material [100] [101] [102]. Metal stents are nonexpendable tubular stable (non-degradable) where polymeric stents were tried as internal stents with tracheomalacia [103] [104]. The results from these studies suggest that stenting is a promising method for treating tracheal obstruction.

2.4.3 Stents in the Esophagus and Gastrointestinal Tract

Many malignant and benign esophageal and gastrointestinal strictures can be treated by minimally invasive alternatives to surgery, including the use of stents. The most commonly used stents in the esophagus and in gastrointestinal tract are the Esophacoil, Flamingo, Gianturco-Z, Ultraflex and Wallstent stents. These stents are generally effective in relieving esophageal dysphasia [105] [106] [107], a success that has led to the employment of stents to manage lesions of the gastrointestinal tract, including the stomach, pylorus, duodenum, upper small intestine and colon [106] [107]. Use of bioresorbable materials for the esophageal stent is currently being explored.

2.5 Theoretical Models to Describe Drug Releasing Behavior

Biodegradable polymeric coatings on cardiovascular stents can be used for local delivery of therapeutic agents to diseased coronary arteries after stenting procedures. A valid mathematical model can be a very important tool in the design and development of such coating for drug delivery. The model should incorporate the important physicochemical processes responsible for the polymer degradation and drug release. Such a model can be used to study the effect of different coating parameters and configurations on the degradation and the release of the drug from the coating. A simultaneous transport-reaction model predicting the degradation and release of the drug Everolimus from a poly lactic acid (PLA) based stent coating has been modeled which describe transportation of water into polymeric matrix, transport of degraded PLA monomers, oligomers, lactic acid and drug using second order partial differential equation with respect to time and radial position [108].These differential equations were then solved using numerical methods. The previously mentioned model was further modified using by assuming two different phase of drug distribution: highly percolated phase and polymer encapsulated phase [109]. Another theoretical model is found which characterize drug release behavior of drug-eluting stents with durable polymer matrix coating [110]. In this model, in vitro and in vivo drug release and tissue pharmacokinetics was described by a second order partial differential equation of drug concentration which was then solved by analytical procedure using Bessel function.

2.6 Conclusion

From the review it is found that Coronary artery disease is the most common type of heart diseases and the leading cause of death worldwide due to heart disease. The drug-eluting stents is generally considered as the best catheter-based therapy for coronary artery disease. However, the mass transfer of the drug from DES to the tissue is not fully understood till date.

Chapter Three

OBJECTTIVE WITH SPECIFIC AIMS AND RESEARCH SIGNIFICANCE

3.1 Objectives of Study with Specific Aims

Coronary artery disease is the most common type of heart diseases and the leading cause of death worldwide due to heart disease. The drug-eluting stents is generally considered as the best catheter-based therapy for coronary artery disease. However, the mass transfer of the drug from DES to the tissue is not fully understood till date. Hence, the main objective of this thesis is to simulate the overall mass transfer phenomenon occurs in CAD-DES in the treatment of coronary artery disease. When a biodegradable coating (containing drug at the porous place of the coating) is given on the main structure of the stent, then this coating is degraded with time and drug releasing property from the stent also become changed. With degraded coating, the diffusion resistance of the drug is changed. At the same time the characteristic diffusion length and drug concentration also become changed. Thus, the thesis work focus on the change in coating resistance as well as the equilibrium drug concentration with respect to time and drug concentration changing due to changed equilibrium drug concentration with respect to time and position (radial) in the artery wall. These changes are observed by solving two partial differential equations for concentration and coating thickness where initial and boundary conditions are assumed. The simulated model is intended for the following specific aims:

- To show the decay of drug coating thickness (including the change in concentration of therapeutic drug in the coating) with time.
- To show the change in drug concentration in the artery tissue layer with time both in radial and longitudinal directions (i.e., a two-dimensional concentration profile).
- The model will finally be generalized incorporating an extra-component of drug mass transfer from the drug coating through the bare stent into the external blood flow in artery channel.

3.2 Research Significance

The main target of this research work was to develop a model which will be able to predict drug mass transfer in CAD-DES. The significance or importance of this model has been briefly indicated below:

➤ The time taken by drug coating to decay completely can reasonably be determined using the simulated model. This will help to determine the safe working period of a drug eluting stent (DES). The model can be employed to study the effect of the properties of DES on their safe working period.
➤ The distribution of therapeutic drug molecules in the artery tissue along axial and radial directions after the release from stent can be studied using the simulated model. This will greatly help to evaluate the therapeutic performances of the various DES.
➤ The time taken by therapeutic drug molecules to cover the entire artery tissue can also reasonably be determined from the simulated model. This will help to evaluate the activity fastness of a DES.
➤ The simulated model can also be used to optimize several parameters related to DES and drug concentration profile.

Overall, the design, operation and performance of DES-based angioplasty would be benefited immensely from the present study.

Chapter Four

RESEARCH METHODOLOGY

Mass transport refers to the movement of mass, i.e. the species of interest which is drugs in the case of a DES, within a defined system. This transport of species may be provoked by concentration gradients between two points, but quite often in systems, especially in the vasculature, overpowering complex flow dynamics will ultimately be responsible for the mass transport outcome. In the absence of a free flowing system the presence of these concentration gradients induces diffusion, e.g. between the DES and the artery wall. Mass transport can be broken up into two types within the human vasculature. Firstly blood side mass transport (BSMT) refers to species transport within the vessel lumen and is subject to the hemodynamic therein. Often evanescent due to hemodynamic washout, BSMT can only be effective in transporting anti-proliferative agents to the wall in regions of high recirculation. The second, and most important, mode of mass transport is in relation to transport within the wall of the artery, referred to as wall side mass transport (WSMT). Along with the properties of the species being transported within the artery wall, WSMT depends on the structural condition of the wall itself, whereby a damaged intimal layer could facilitate accelerated mass transport through to the medial layer. WSMT can be governed by two transport forces, a pressure driven convective force and a diffusive force. The Peclet number (Pe) is a dimensionless parameter that can be used to determine the relative influences of these two forces. A small Pe (Pe<1) is representative of transport which is dominated by diffusion, while a higher Pe (Pe>1) indicates convection dominated mass transport.

4.1 Governing Equation of the Problem

In the present thesis work, the main objective was to consider WSMT where drug concentration was evaluated with respect to time and position in artery wall. In this purpose, two drug mass transfer model was developed. In 1D model drug concentration has been evaluated with respect to time and radial position (r) in the artery wall. In 2D model a drug concentration profile was developed with respect to time and radial (r) and longitudinal position (z) in the artery wall.

The assumptions applied when modeling fluid flow problems of DES are as follows:

- The flow is incompressible and isothermal
- The fluid is Newtonian and possesses constant physical properties
- Flow is considered to be laminar
- Drug coating over the DES was a continuous film

Species transport via diffusion is a process driven by concentration gradients between two locations. Fick's first law can be used to describe the diffusion flux (Jx, mol/m²s) of such species, shown in 1D in equation 4.1, where D (m²/s) is diffusivity and c is concentration (mol/m³):

$$J_x = -D \frac{\partial c}{\partial x} \qquad (4.1)$$

The negative term in equation 4.1 indicates that the flux is positive in the presence of a negative concentration gradient. Biological mass transport often requires the application of a time-dependent mass transport process that can predict variations in concentration overtime. Fick's second law (equation 4.2) can provide such a relationship and is defined here in one dimension:

$$\frac{\partial c}{\partial t} = D \frac{\partial^2 c}{\partial x^2} \qquad (4.2)$$

The addition of a convective term, equal to the product of the fluid velocity and the local concentration, to equation 4.2 demonstrates the 3D transport of species in a flowing solution. This is known as the convection-diffusion equation.

$$\frac{\partial c}{\partial t} + u \frac{\partial c}{\partial x} + v \frac{\partial c}{\partial y} + w \frac{\partial c}{\partial z} = D \left(\frac{\partial^2 c}{\partial x^2} + \frac{\partial^2 c}{\partial y^2} + \frac{\partial^2 c}{\partial z^2} \right) \qquad (4.3)$$

For a cylindrical coordinate system, the equation 4.3 becomes,

$$\frac{\partial c}{\partial t} + u \frac{\partial c}{\partial r} + \frac{v}{r} \frac{\partial c}{\partial \theta} + w \frac{\partial c}{\partial z} = D \left[\frac{1}{r} \frac{\partial}{\partial r} \left(r \frac{\partial c}{\partial r} \right) + \frac{1}{r^2} \frac{\partial^2 c}{\partial \theta^2} + \frac{\partial^2 c}{\partial z^2} \right] \qquad (4.4)$$

Here, u, v and w are the velocity at the r, θ and z direction respectively.

In reality the classification of problems of this nature are inherently patient specific and as such no one representation of the problem is correct. However, there are innate similarities between patients. Blood flow within the vasculature is a highly complex 3D process to model given the pulsatile nature of arterial haemo-dynamics. Coupled with this pulsatile process, the coronary arteries are situated on the surface of the heart and as such are subject to cyclic motion due to the beating of the organ. Therefore the modeling of drug transport from a DES in these arteries is multifaceted in nature, comprising of both luminal and artery wall mass transport, the latter of which may also be subject to a reaction giving that some drug may bind to the arterial tissue. The introduction of a multi-layered artery wall to the model increases the complexity of the domain even further. If it is considered that, DES has been placed in multilayer diseased artery, then in absence of a lumen and subsequent blood stream mass transport, the wall side mass transport is approximately diffusive with a little convection flux arisen due to high concentration difference. Thus the diffusion of drug from DES occurs at the radial direction with a small amount convection of drug at the r direction with a velocity u. In equation 4.4, setting velocity v and w to zero and concentration gradient at θ and z direction to zero, it is found that,

$$\frac{\partial c}{\partial t} + u\frac{\partial c}{\partial r} = D\frac{1}{r}\frac{\partial}{\partial r}\left(r\frac{\partial c}{\partial r}\right)$$

$$\Rightarrow \frac{\partial c}{\partial t} = D\frac{\partial^2 c}{\partial r^2} + \frac{D}{r}\frac{\partial c}{\partial r} - u\frac{\partial c}{\partial r}$$

$$\frac{D}{r} \to 0 \; asD \ll r; \; \frac{\partial c}{\partial t} = D\frac{\partial^2 c}{\partial r^2} - u\frac{\partial c}{\partial r} \qquad (4.5)$$

Now by putting, $t = \frac{R^2 \bar{t}}{D}$; $r = \bar{r}R$ and $c = \bar{c}c_0$ equation 4.5 can be converted into dimensionless format where, $\bar{t}, \bar{r}$ and $\bar{c}$ are dimensionless time, radius and concentration respectively.

$$\frac{\partial(\bar{c}c_0)}{\partial\left(\frac{R^2\bar{t}}{D}\right)} = D\frac{\partial^2(\bar{c}c_0)}{\partial(\bar{r}R)^2} - u\frac{\partial(\bar{c}c_0)}{\partial(\bar{r}R)}$$

$$\Rightarrow \frac{Dc_0}{R^2}\frac{\partial \bar{c}}{\partial \bar{t}} = \frac{Dc_0}{R^2}\frac{\partial^2 \bar{c}}{\partial \bar{r}^2} - \frac{uc_0}{R}\frac{\partial \bar{c}}{\partial \bar{r}}$$

Dividing both side by, $\frac{Dc_0}{R^2}$, the equation become

$$\frac{\partial \bar{c}}{\partial \bar{t}} = \frac{\partial^2 \bar{c}}{\partial \bar{r}^2} - \frac{uR}{D}\frac{\partial \bar{c}}{\partial \bar{r}}$$

$$\Rightarrow \frac{\partial \bar{c}}{\partial \bar{t}} = \frac{\partial^2 \bar{c}}{\partial \bar{r}^2} - Pe\frac{\partial \bar{c}}{\partial \bar{r}} \tag{4.6}$$

Equation 4.6 is the governing equation of the drug concentration in artery wall at dimensionless form, where unsteady drug concentration is considered in radial direction only.

A common assumption for DES mass transport studies is that the intimal layers of the artery are denuded and that the stent is in direct contact with the medial layer of the artery wall. This negates the need to model the endothelial, intima and internal elastic lamina layers. Regardless of the inclusion or exclusion of these layers the continuity equation should be the default setting for all interior boundaries. This condition states that in the absence of sources or sinks, the flux in the normal direction is continuous across the boundary, i.e. the concentration is equal on both sides of the boundary

At the vascular wall and at the up- and down-stream wall boundaries there should be a sufficient distance away from the stent. It specifies where the domain is well insulated or it can reduce the size of a model by taking advantage of symmetry. Intuitively this condition states that the gradient across the boundary must be zero, therefore it is impermeable to mass transport.

Again if, diffusion of drug from DES occurs at the radial direction with a small amount diffusion of drug at the z direction due to high concentration gradient, then from equation 4.4 it is found that,

$$\frac{\partial c}{\partial t} + u\frac{\partial c}{\partial r} = D\frac{1}{r}\frac{\partial}{\partial r}\left(r\frac{\partial c}{\partial r}\right) + D\frac{\partial^2 c}{\partial z^2} \tag{4.7}$$

Equation 4.7 is the governing equation of the drug concentration in artery wall at radial and axial direction with respect to time.

The implementation of an arterial pulse and a beating heart are neglected by most researchers. Often the artery is modeled as rigid in space in order to analyze mass transport post DES deployment. This is an effective assumption but one must consider the deformation of the artery wall due to the dynamic expansion of the stent, as this can have an impact on the mass transport outcome due to the porous nature of the wall and the compression it incurs upon stent expansion. As for the application of laminar blood flow, it can be seen that the majority of drug that enters the artery wall from the DES does so via physical contact with the wall and the drugs emanating from the areas of the stent exposed to flow, be it laminar or pulsatile, are predominantly carried downstream.

In this Thesis work, the coating given over the metal stent of the DES was considered as biodegradable. Thus reduction of the coating thickness was described by the following equation

$$\frac{\partial h}{\partial t} = -kh^n \qquad (4.8)$$

Where, k and n are two constant which are optimized according to property of coating polymer and drug release time.

Boundary condition of equation 4.8 is: At $t = 0$, $h = h_0$

If an extra component of drug mass transfer occurs from the drug coating through the bare stent into the external blood flow in artery channel, then two different concentration terms have to be considered one is WSMT or wall side mass transfer and another is BSMT or blood side mass transfer. WSMT is almost diffusive with very little convection and BSMT is mostly convective with little diffusion in general. Thus for WSMT equation 4.7 dives the concentration where for BSMT, from equation 4.4 it is found that

$$\frac{\partial c_b}{\partial t} + w\frac{\partial c_b}{\partial z} = D\frac{1}{r}\frac{\partial}{\partial r}\left(r\frac{\partial c_b}{\partial r}\right)$$

$$\Rightarrow \frac{\partial c_b}{\partial t} = D\frac{1}{r}\frac{\partial}{\partial r}\left(r\frac{\partial c_b}{\partial r}\right) - w\frac{\partial c_b}{\partial z} \qquad (4.9-i)$$

Then incorporating equation 4.7, 4.8 and 4.9-i the total amount of drug release from the DES is found by following equation,

$$\frac{\partial c_T}{\partial t} = \frac{\partial c}{\partial t} + \frac{\partial c_b}{\partial t} \qquad (4.9-ii)$$

To get the concentration profile of drug in artery wall, equation 4.6 and 4.7 is required to be solved. Both of the equations are partial differential equation. An analytical solution of equation 4.6 might be possible using Bassel function as it is a second order partial differential equation of two unknown variable, but the solution would contain many unknown terms. Equation 4.7 is a second order partial differential equation containing three independent variables, which has no direct analytical solution. Thus computational method (finite volume algorithm) has been selected to solve these equations. Simulation procedure of these equations are given in section 4.3.

4.1.1 Dimensionless Parameter of Drug Release

The Peclet number (Pe) is a dimensionless number that determines the relative contribution of convective and diffusive forces to species transport within a defined system. It can be defined as a product of the Reynolds number (Re) and the Schmidt number (Sc).

$$Pe = Re.Sc \qquad (4.10)$$

The Reynolds number is a non-dimensional parameter concerning fluid forces due to viscosity and inertia and is essentially used to determine whether a flow is laminar, transitional or turbulent in nature. For example a Reynolds number of approximately 90 can be obtained for a mean arterial velocity (u) of 0.1m/s in an artery with a diameter (a) of 3mm. When considering trans-mural flow through the porous artery wall the value a would represent the thickness of the porous wall.

$$Re = \frac{\rho u a}{\mu} \qquad (4.11)$$

The Schmidt number (Sc) is defined as the ratio of Kinematic viscosity (υ, m²/s) to diffusivity (D)

$$Sc = \frac{\upsilon}{D} = \frac{\mu}{\rho D} \qquad (4.12)$$

Substituting equation 4.11 and 4.12 into 4.10 describes how convective and diffusive forces can influence the outcome of the Peclet number.

$$Pe = \frac{\rho u a}{\mu} \times \frac{\mu}{\rho D} = \frac{u a}{D} \qquad (4.13)$$

4.1.2 Diffusion in Porous Material

When considering diffusion in a fluid saturated porous media, as is the case with the artery wall, diffusion takes place over a tortuous path. Because these pores are not straight, the distance over which diffusion takes place becomes effectively longer than for a homogenous material of the same thickness. The effective diffusivity (D_{eff}) can therefore be deduced by considering the impact of the materials structure on the species free diffusivity (D_{free}). The effective diffusivity of a porous material is a function of its porosity (ε) and tortuosity (τ).

$$D_{eff} = \frac{\varepsilon}{\tau} D_{free} \qquad (4.14)$$

One of the more common ways to determine the free diffusivity of a species in a solvent is to use the stokes-Einstein equation (4.15), where k (J/K) is the Boltzmann constant, T (K) is the temperature and R (m) is the radius of the diffusing molecule. For the purpose of diffusion in the artery wall, the wall layer is considered to be at plasma state.

$$D_{free} = \frac{kT}{6\pi\mu R} \qquad (4.15)$$

The radius of the solute can be calculated from equation 4.16 assuming that the particle is spherical in shape, M (kg/mol) is the drug molecular weight and Na (mol^{-1}) is the Avogadro's number.

$$R = \left(\frac{3M}{4\pi\rho Na}\right)^{\frac{1}{3}} \qquad (4.16)$$

The structure of the porous medium is defined by the tortuosity (τ) of its porous network (4.17) and by the porosity (ε) (4.18) of the material itself.

$$\tau = \frac{L}{X} \qquad (4.17)$$

Where L = pore path length and X = distance between beginning and end of the pore path.

$$\varepsilon = \frac{Pore\, volume}{Total\, volume} \qquad (4.18)$$

4.1.3 Artery Wall Classification

Arteries transport oxygen rich blood around the body providing essential nutrients to vital organs. The artery wall consists of a complex multilayer porous substructure with an interstitial phase comprising predominantly of plasma. In a healthy artery this substructure(Fig.4.1) is comprised of three concentric layers; the tunica intima, the tunica media and the tunica adventitia. The tunica intima is the innermost layer, consisting of a single layer of endothelial cells and a sub endothelial layer mainly consisting of delicate connective tissues and collagen fibers. The outer boundary of the tunica intima is surrounded by an elastic tissue with fenestral pores known as the internal elastic lamina (IEL). The medial layer consists primarily of concentric sheets of smooth muscle cells (SMC) within a loose connective tissue framework. This configuration of SMC enables the artery wall to contract and relax. The tunica media and the tunica adventitia are separated by another thin band of elasticfibers known as the external elastic lamina (EEL). The outermost layer of the artery, the tunica adventitia, is comprised of connective tissue fibers and some capillaries. These fibers blend into the surrounding connective tissues and aid in stabilizing the arteries within the body. The target layer for the anti-restenotic drugs is the tunica media, because SMC resides here and possible erosion of the tunica intima occurs upon stent deployment.

The artery wall is porous in composition and drug transport is facilitated through the surrounding plasma not only via diffusion but there is also the presence of a trans-mural velocity due to a pressure gradient observed across the artery wall. However, the presence of arterial plaque will reduce the magnitude of this trans-mural velocity and can even stem it altogether. As DES is deployed in highly occluded arteries it is reasonable to reduce the complexity of the problem by neglecting convection in the wall. Equation 4.14 gives us an indication of how arterial properties such as porosity, tortuosity and free diffusivity can influence the transport of drugs within the respective artery wall layers. The compression of these layers will alter these properties which in turn may inhibit the transport of species as governed by the mass transport equations. The compression of a porous structure not only reduces the materials porosity but it results in the creation of a more arduous pore path over which mass transport would normally occur. The combination of a reduced porosity with an increased tortuosity, when the artery wall has been compressed, has a net effect of reducing the effective diffusivity thus hindering mass transport within the vessel.

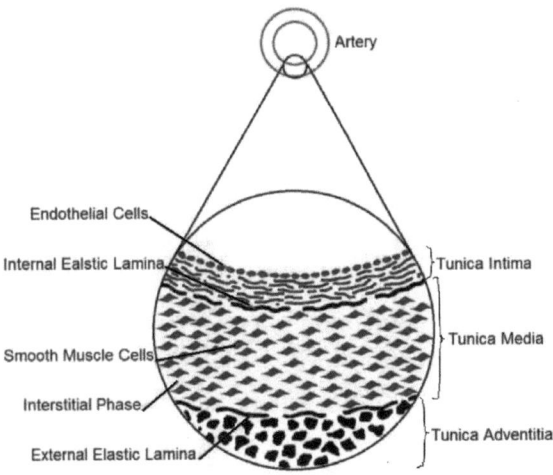

Fig. 4.1: Illustration of the cross sectional structure of a healthy artery wall

4.2 Computational Fluid Dynamics

Computational fluid dynamics or CFD is the analysis of the system involving fluid flow, heat transfer and associated phenomenon such as chemical reactions by means of computer-based simulation. The technique is very powerful and spans a wide range of industrial and non-industrial application areas. It has emerged as one of the most powerful numerical tools for engineers, scientists and mathematicians alike. Its foundations are based on theoretical analysis drawn from experimental observations over various branches of physics.

There are several unique advantages of CFD over the experiment based approaches to the fluid system design. Some of those are:

- Substantial reduction of lead time and cost of new designs
- Ability to study system where controlled experiments are difficult or impossible to perform
- Ability to study systems under hazardous conditions at and beyond their normal performance limit
- Practically unlimited level of detail of results

The starting point for any computational analysis is the appropriate allocation of the governing equations. These equations are then substituted with equivalent numerical descriptions that are then solved using appropriate mathematical techniques. There are a number of numerical techniques available that will return a solution to a specified problem. Two of the more popular methods are the Finite Volume Method and the Finite Element Method. Three mathematical concepts are useful to determine the success of such methods: convergence, consistency and stability.

Convergence is the property of a numerical method to produce a solution which approaches the exact solution as the grid spacing; control volume size is reduced to zero. Consistent numerical schemes produce system of algebraic equations which can be demonstrated to be equivalent to the original governing equations the grid spacing tends to zero. Stability is associated with damping of errors as the numerical method proceeds. If a technique is not stable even round-off errors in the initial data can cause wild oscillations or divergence.

4.2.1 Finite Volume Method

Finite volume method was originally developed as a special finite difference formulation. It is one of the most well-established and thoroughly validated general purpose CFD techniques. It is the central to four of the five main commercially available CFD codes: PHOENICS, FLUENT, FLOW3D and STAR-CD. This method consists of the following steps:

- ➢ Formal integration of the governing equations of the fluid flow over all the control volumes of the solution domain
- ➢ Discretization involves the substitution of a variety of finite difference type approximations for the terms in the integrated equation representing flow processes such as convection, diffusion and sources. This converts the integral equations into a system of algebraic equations
- ➢ Solution of the algebraic equations by an iterative methods

The first step, the control volume integration, distinguishes the finite volume method from all other CFD techniques. The resulting statements express the conservation of relevant properties for each finite size cells. This clear relationship between the numerical algorithm and the underlying physical conservation principle forms one of the main attractions of the finite volume methods and make its concepts much simpler to understand. For a certain variable Ø, the general conservation equation of the finite volume method is:

$$\begin{bmatrix} Rate\ of\ change \\ of\ \emptyset\ in\ control \\ volume\ with \\ respect\ to\ time \end{bmatrix} = \begin{bmatrix} Net\ flux\ of \\ \emptyset\ due\ to \\ convection\ into \\ control\ volume \end{bmatrix} + \begin{bmatrix} Net\ flux\ of \\ \emptyset\ due\ to \\ diffusion\ into \\ control\ volume \end{bmatrix} + \begin{bmatrix} Net\ rate\ of \\ creation\ of\ \emptyset \\ inside\ the \\ control \\ volume \end{bmatrix}$$

CFD codes for finite volume technique contains discretization techniques suitable for the treatment of the key transport phenomenon, convection and diffusion as well as source terms and the rate of change with respect to time. The underlying physical phenomenon is complex a non-linear so an iterative solution approach is required.

Finite approach guarantees local conservation of a fluid property φ for each control volume. Numerical schemes which possess the conservativeness property also ensure global conservation of the fluid property for the entire domain. This is clearly important physically and is achieved by means of consistent expressions for fluxes of φ through the cell faces of the adjacent volumes. Further detail of this method has been given in Appendix A.

4.3 Simulation of the Governing Equation

4.3.1 Solution of 1D Unsteady Concentration Equation

Using finite volume approach for solving the concentration partial differential equation, finite volume element is considered of volume ΔV within face point e and w.

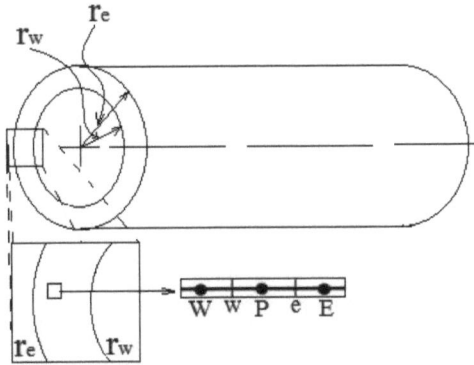

Fig. 4.2: 1D Finite volume in cylindrical system

Here, $\Delta V = A dr$. In fig.4.2 point P is the node where, concentration will be considered within the volume. W and e indicate the faces of the finite volume. W and E are the node point of the adjacent finite volume. Now, equation 4.6 is integrated twice within the finite ΔV and time limit t to t + Δt.

$$\int_t^{t+\Delta t}\left[\int_{\Delta V}\frac{\partial \bar{c}}{\partial \bar{t}}dV\right]d\bar{t} = \int_t^{t+\Delta t}\left[\int_{\Delta V}\frac{\partial^2 \bar{c}}{\partial \bar{r}^2}dV\right]d\bar{t} - Pe\int_t^{t+\Delta t}\left[\int_{\Delta V}\frac{\partial \bar{c}}{\partial \bar{r}}dV\right]d\bar{t}$$

$$\Rightarrow \int_t^{t+\Delta t}\frac{\partial \bar{c}}{\partial \bar{t}}\Delta V d\bar{t} = \int_t^{t+\Delta t}\int_w^e \frac{\partial}{\partial \bar{r}}\left(\frac{\partial \bar{c}}{\partial \bar{r}}\right)A d\bar{r}d\bar{t} - Pe\int_t^{t+\Delta t}\int_w^e \frac{\partial \bar{c}}{\partial \bar{r}}A d\bar{r}d\bar{t}$$

$$\Rightarrow \frac{\bar{c}-\bar{c}^0}{\Delta \bar{t}}\Delta V \Delta \bar{t} = \int_t^{t+\Delta t}\int_w^e \partial\left(\frac{\partial \bar{c}}{\partial \bar{r}}\right)A d\bar{t} - Pe\int_t^{t+\Delta t}\int_w^e \partial \bar{c}A d\bar{t}$$

$$\Rightarrow A_P \Delta \bar{r}(\bar{c}-\bar{c}^0) = \int_t^{t+\Delta t}\left[A_e\left(\frac{\partial \bar{c}}{\partial \bar{r}}\right)_e - A_w\left(\frac{\partial \bar{c}}{\partial \bar{r}}\right)_w\right]d\bar{t} - Pe\int_t^{t+\Delta t}[A_e \bar{c}_e - A_w \bar{c}_w]d\bar{t}$$

Integrating time integral with θ time fraction, where θ fraction of previous time step and (1 − θ) time fraction of the current time step is taken into account.

$$\Rightarrow A_P \Delta \bar{r}(\bar{c}_P - \bar{c}_P^0)$$
$$= \left[A_e\frac{\bar{c}_E - \bar{c}_P}{\Delta \bar{r}} - A_w\frac{\bar{c}_P - \bar{c}_W}{\Delta \bar{r}}\right]^\theta \Delta \bar{t}$$
$$- Pe\left[A_e\frac{\bar{c}_E + \bar{c}_P}{2} - A_w\frac{\bar{c}_P + \bar{c}_W}{2}\right]^\theta \Delta \bar{t} \quad (4.19)$$

$$\Rightarrow \frac{A_P \Delta \bar{r}}{\Delta \bar{t}}(\bar{c}_P - \bar{c}_P^0)$$
$$= \theta\left[\frac{A_e}{\Delta \bar{r}}\bar{c}_E + \frac{A_w}{\Delta \bar{r}}\bar{c}_W - \left(\frac{A_e}{\Delta \bar{r}} + \frac{A_w}{\Delta \bar{r}}\right)\bar{c}_P\right]$$
$$+ (1-\theta)\left[\frac{A_e}{\Delta \bar{r}}\bar{c}_E^0 + \frac{A_w}{\Delta \bar{r}}\bar{c}_W^0 - \left(\frac{A_e}{\Delta \bar{r}} + \frac{A_w}{\Delta \bar{r}}\right)\bar{c}_P^0\right]$$
$$- \frac{Pe}{2}\theta[A_e\bar{c}_E - A_w\bar{c}_W + (A_e - A_w)\bar{c}_P]$$
$$- \frac{Pe}{2}(1-\theta)[A_e\bar{c}_E^0 - A_w\bar{c}_W^0 + (A_e - A_w)\bar{c}_P^0]$$

$$\Rightarrow \frac{A_P \Delta \bar{r}}{\Delta \bar{t}} \bar{c}_P - \frac{A_P \Delta \bar{r}}{\Delta \bar{t}} \bar{c}_P^0$$

$$= \theta A_e \left[\frac{1}{\Delta \bar{r}} - \frac{Pe}{2}\right] \bar{c}_E + \theta A_w \left[\frac{1}{\Delta \bar{r}} + \frac{Pe}{2}\right] \bar{c}_W$$

$$- \theta \left[\left(\frac{A_e}{\Delta \bar{r}} + \frac{A_w}{\Delta \bar{r}}\right) + \frac{Pe}{2}(A_e - A_w)\right] \bar{c}_P + (1-\theta) A_e \left[\frac{1}{\Delta \bar{r}} - \frac{Pe}{2}\right] \bar{c}_E^0$$

$$+ (1-\theta) A_w \left[\frac{1}{\Delta \bar{r}} + \frac{Pe}{2}\right] \bar{c}_W^0 - (1-\theta) \left[\left(\frac{A_e}{\Delta \bar{r}} + \frac{A_w}{\Delta \bar{r}}\right) + \frac{Pe}{2}(A_e - A_w)\right] \bar{c}_P^0$$

$$\Rightarrow -\theta A_e \left[\frac{1}{\Delta \bar{r}} - \frac{Pe}{2}\right] \bar{c}_E + \left[\frac{A_P \Delta \bar{r}}{\Delta \bar{t}} + \theta \left\{\left(\frac{A_e}{\Delta \bar{r}} + \frac{A_w}{\Delta \bar{r}}\right) + \frac{Pe}{2}(A_e - A_w)\right\}\right] \bar{c}_P$$

$$- \theta A_w \left[\frac{1}{\Delta \bar{r}} + \frac{Pe}{2}\right] \bar{c}_W$$

$$= (1-\theta) A_e \left[\frac{1}{\Delta \bar{r}} - \frac{Pe}{2}\right] \bar{c}_E^0 + (1-\theta) A_w \left[\frac{1}{\Delta \bar{r}} + \frac{Pe}{2}\right] \bar{c}_W^0$$

$$- \left[(1-\theta)\left\{\left(\frac{A_e}{\Delta \bar{r}} + \frac{A_w}{\Delta \bar{r}}\right) + \frac{Pe}{2}(A_e - A_w)\right\} - \frac{A_P \Delta \bar{r}}{\Delta \bar{t}}\right] \bar{c}_P^0 \quad (4.20)$$

Equation 4.20 is the algebraic form of the governing equation. For different time steps and for different radial positions, this equation can be solved. Now for getting better result, taking implicit method of problem solving, θ is set to 1 and thus equation 4.20 become,

$$-A_e \left[\frac{1}{\Delta \bar{r}} - \frac{Pe}{2}\right] \bar{c}_E + \left[\frac{A_P \Delta \bar{r}}{\Delta \bar{t}} + \left(\frac{A_e}{\Delta \bar{r}} + \frac{A_w}{\Delta \bar{r}}\right) + \frac{Pe}{2}(A_e - A_w)\right] \bar{c}_P - A_w \left[\frac{1}{\Delta \bar{r}} + \frac{Pe}{2}\right] \bar{c}_W$$

$$= \frac{A_P \Delta \bar{r}}{\Delta \bar{t}} \bar{c}_P^0 \quad (4.21)$$

This is the final algebraic equation of finite volume method, which can be written in following form

$$-a_E \bar{c}_E + a_P \bar{c}_P - a_W \bar{c}_W = a_P^0 \bar{c}_P^0 \quad (4.22)$$

Comparing equation 4.21 and 4.22, the coefficient of the concentration terms can be written as follows where subscript n indicates discretized point from 2 to N – 1.

$$a_E = A_e \left[\frac{1}{\Delta \bar{r}} - \frac{Pe}{2}\right] = \frac{A_P + A_E}{2} \left[\frac{1}{\Delta \bar{r}} - \frac{Pe}{2}\right] = \frac{2\pi L(\bar{r}_P + \bar{r}_E)}{2} \left[\frac{1}{\Delta \bar{r}} - \frac{Pe}{2}\right]$$

$$\Rightarrow a_{En} = \pi L(\bar{r}_P + \bar{r}_E) \left[\frac{1}{\Delta \bar{r}} - \frac{Pe}{2}\right] \quad (4.23)$$

$$a_W = A_w \left[\frac{1}{\Delta \bar{r}} + \frac{Pe}{2}\right] = \frac{A_P + A_W}{2}\left[\frac{1}{\Delta \bar{r}} + \frac{Pe}{2}\right] = \frac{2\pi L(\bar{r}_P + \bar{r}_W)}{2}\left[\frac{1}{\Delta \bar{r}} + \frac{Pe}{2}\right]$$

$$\Rightarrow a_{Wn} = \pi L(\bar{r}_P + \bar{r}_W)\left[\frac{1}{\Delta \bar{r}} + \frac{Pe}{2}\right] \tag{4.24}$$

$$a_{Pn}^0 = \frac{A_P \Delta \bar{r}}{\Delta \bar{t}} = \frac{2\pi L \bar{r}_P \Delta \bar{r}}{\Delta \bar{t}}$$

$$a_P = \frac{A_P \Delta \bar{r}}{\Delta \bar{t}} + \left(\frac{A_e}{\Delta \bar{r}} + \frac{A_w}{\Delta \bar{r}}\right) + \frac{Pe}{2}(A_e - A_w) = A_e\left[\frac{1}{\Delta \bar{r}} - \frac{Pe}{2}\right] + A_w\left[\frac{1}{\Delta \bar{r}} + \frac{Pe}{2}\right] + \frac{A_P \Delta \bar{r}}{\Delta \bar{t}}$$

$$\Rightarrow a_{Pn} = a_{En} + a_{Wn} + a_{Pn}^0 \tag{4.25}$$

Element of column matrix, $b_n = a_{Pn}^0 \bar{c}_{Pn}^0$ \hfill (4.26)

In matrix A, the diagonal component of 1st and the last row are a_{P1}, a_{E1}, a_{PN} and a_{WN} are calculated separately because, when node point 1 is considered as present point, there is no west adjacent point at the adjacent as it indicates the point on inner wall of the artery. Again, during considering node point N as present point, there is no east adjacent point as point N is situated on the on the outer wall of the artery.

At the west or the inner wall of artery, from equation 4.19 it is found that

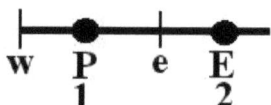

Fig. 4.3: Inner wall face boundary and node point of discretized volume

$$A_P \Delta \bar{r}(\bar{c}_P - \bar{c}_P^0) = \left[A_e \frac{\bar{c}_E - \bar{c}_P}{\Delta \bar{r}} - A_w \frac{\bar{c}_P - \bar{c}_w}{\frac{\Delta \bar{r}}{2}}\right]^\theta \Delta \bar{t} - Pe\left[A_e \frac{\bar{c}_E + \bar{c}_P}{2} - A_w \bar{c}_w\right]^\theta \Delta \bar{t}$$

Considering implicit scheme by putting θ = 1

$$\Rightarrow A_P(\bar{c}_P - \bar{c}_P^0)\frac{\Delta \bar{r}}{\Delta \bar{t}} = A_e \frac{\bar{c}_E - \bar{c}_P}{\Delta \bar{r}} - 2A_w \frac{\bar{c}_P - \bar{c}_w}{\Delta \bar{r}} - PeA_e \frac{\bar{c}_E + \bar{c}_P}{2} + PeA_w \bar{c}_w$$

$$\Rightarrow \left(A_P \frac{\Delta \bar{r}}{\Delta \bar{t}} + \frac{A_e}{\Delta \bar{r}} + \frac{2A_w}{\Delta \bar{r}} + \frac{PeA_e}{2}\right) \bar{c}_P - \left(\frac{A_e}{\Delta \bar{r}} - \frac{PeA_e}{2}\right) \bar{c}_E$$

$$= A_P \frac{\Delta \bar{r}}{\Delta \bar{t}} \bar{c}_P^0 + \left(\frac{2A_w}{\Delta \bar{r}} + PeA_w\right) \bar{c}_w \quad (4.27)$$

This equation can be written as

$$a_{P1}\bar{c}_{P1} - a_{E1}\bar{c}_E = a_{P1}^0 \bar{c}_{P1}^0 + a_w \bar{c}_w$$

Comparing both equations,

$$a_{E1} = \frac{A_e}{\Delta \bar{r}} - \frac{PeA_e}{2} = \frac{A_P + A_E}{2}\left[\frac{1}{\Delta \bar{r}} - \frac{Pe}{2}\right] = \frac{2\pi L(\bar{r}_P + \bar{r}_E)}{2}\left[\frac{1}{\Delta \bar{r}} - \frac{Pe}{2}\right]$$

$$\Rightarrow a_{E1} = \pi L (\bar{r}_P + \bar{r}_E) \left[\frac{1}{\Delta \bar{r}} - \frac{Pe}{2}\right] \quad (4.28)$$

$$a_w = \frac{2A_w}{\Delta \bar{r}} + PeA_w = 2\pi \bar{r}_w L \left(\frac{2}{\Delta \bar{r}} + Pe\right)$$

$$a_{P1}^0 = A_P \frac{\Delta \bar{r}}{\Delta \bar{t}} = 2\pi \bar{r}_P L \frac{\Delta \bar{r}}{\Delta \bar{t}}$$

$$a_{P1} = A_P \frac{\Delta \bar{r}}{\Delta \bar{t}} + \frac{A_e}{\Delta \bar{r}} + \frac{2A_w}{\Delta \bar{r}} + \frac{PeA_e}{2} = A_P \frac{\Delta \bar{r}}{\Delta \bar{t}} + \frac{A_P + A_E}{2}\left(\frac{1}{\Delta \bar{r}} + \frac{Pe}{2}\right) + \frac{2A_w}{\Delta \bar{r}}$$

$$\Rightarrow a_{P1} = 2\pi L \left[\bar{r}_P \frac{\Delta \bar{r}}{\Delta \bar{t}} + \frac{(\bar{r}_P + \bar{r}_E)}{2}\left(\frac{1}{\Delta \bar{r}} + \frac{Pe}{2}\right) + \frac{2\bar{r}_w}{\Delta \bar{r}}\right] \quad (4.29)$$

And 1st element of the column matrix B is

$$b_1 = a_{P1}^0 \bar{c}_{P1}^0 + a_w \bar{c}_w \quad (4.30)$$

At the east or the outer wall of artery, from equation 4.19 it is found that

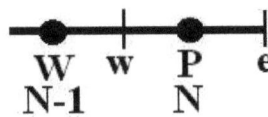

Fig. 4.4: Outer wall face boundary and node point of discretized volume

$$A_P \Delta \bar{r}(\bar{c}_P - \bar{c}_P^0) = \left[A_e \frac{\bar{c}_e - \bar{c}_P}{\frac{\Delta \bar{r}}{2}} - A_w \frac{\bar{c}_P - \bar{c}_w}{\Delta \bar{r}}\right]^\theta \Delta \bar{t} - Pe \left[A_e \bar{c}_e - A_w \frac{\bar{c}_P + \bar{c}_w}{2}\right]^\theta \Delta \bar{t}$$

Considering implicit scheme by putting θ = 1

$$\Rightarrow A_P(\bar{c}_P - \bar{c}_P^0)\frac{\Delta \bar{r}}{\Delta \bar{t}} = 2A_e\frac{\bar{c}_e - \bar{c}_P}{\Delta \bar{r}} - A_w\frac{\bar{c}_P - \bar{c}_w}{\Delta \bar{r}} - PeA_e\bar{c}_e + PeA_w\frac{\bar{c}_P + \bar{c}_w}{2}$$

$$\Rightarrow -A_w\left(\frac{1}{\Delta \bar{r}} + \frac{Pe}{2}\right)\bar{c}_w + \left(A_P\frac{\Delta \bar{r}}{\Delta \bar{t}} + \frac{2A_e}{\Delta \bar{r}} + \frac{A_w}{\Delta \bar{r}} - \frac{PeA_w}{2}\right)\bar{c}_P$$

$$= A_P\frac{\Delta \bar{r}}{\Delta \bar{t}}\bar{c}_P^0 + A_e\left(\frac{2}{\Delta \bar{r}} - Pe\right) \quad (4.31)$$

This equation can be written as,

$$-a_{WN}\bar{c}_{WN} + a_{PN}\bar{c}_{PN} = a_{PN}^0\bar{c}_{PN}^0 + a_{eN}\bar{c}_{eN}$$

Comparing these two equations

$$a_{WN} = A_w\left(\frac{1}{\Delta \bar{r}} + \frac{Pe}{2}\right) = \frac{A_w + A_P}{2}\left(\frac{1}{\Delta \bar{r}} + \frac{Pe}{2}\right) = \pi L(\bar{r}_w + \bar{r}_P)\left(\frac{1}{\Delta \bar{r}} + \frac{Pe}{2}\right) \quad (4.32)$$

$$a_{PN}^0 = A_P\frac{\Delta \bar{r}}{\Delta \bar{t}} = 2\pi L\bar{r}_P\frac{\Delta \bar{r}}{\Delta \bar{t}}$$

$$a_{eN} = A_e\left(\frac{2}{\Delta \bar{r}} - Pe\right) = 2\pi L\bar{r}_e\left(\frac{2}{\Delta \bar{r}} - Pe\right)$$

$$a_{PN} = A_P\frac{\Delta \bar{r}}{\Delta \bar{t}} + \frac{2A_e}{\Delta \bar{r}} + A_w\left(\frac{1}{\Delta \bar{r}} - \frac{Pe}{2}\right) = A_P\frac{\Delta \bar{r}}{\Delta \bar{t}} + \frac{2A_e}{\Delta \bar{r}} + \frac{A_w + A_P}{2}\left(\frac{1}{\Delta \bar{r}} - \frac{Pe}{2}\right)$$

$$\Rightarrow a_{PN} = 2\pi L\left[\bar{r}_P\frac{\Delta \bar{r}}{\Delta \bar{t}} + \frac{2\bar{r}_e}{\Delta \bar{r}} + \frac{\bar{r}_w + \bar{r}_P}{2}\left(\frac{1}{\Delta \bar{r}} - \frac{Pe}{2}\right)\right] \quad (4.33)$$

The Nth element of the column matrix B is

$$b_N = a_{PN}^0\bar{c}_{PN}^0 + a_{eN}\bar{c}_{eN} \quad (4.34)$$

Using equation 4.23, 4.24, 4.25, 4.28, 4.29, 4.32 and 4.33 a three diagonal coefficient matrix A can be constructed. The column matrix B can be constructed using equation 4.26, 4.30 and 4.34, where concentration terms indicate the variable matrix X in equation 4.35.

Then the concentration of therapeutic drug in the each grid point of the artery wall can be evaluated by following way

$$AX = B \Rightarrow X = A^{-1}B \quad (4.35)$$

Where,

$$A = \begin{bmatrix} a_{P1} & -a_{E1} & 0 & 0 & 0 & 0 & . & . & 0 & 0 & 0 & 0 \\ -a_{W2} & a_{P2} & -a_{E2} & 0 & 0 & 0 & . & . & 0 & 0 & 0 & 0 \\ 0 & -a_{W3} & a_{P3} & -a_{E3} & 0 & 0 & . & . & 0 & 0 & 0 & 0 \\ 0 & 0 & -a_{W4} & a_{P4} & -a_{E4} & 0 & . & . & 0 & 0 & 0 & 0 \\ . & . & . & . & . & . & . & . & . & . & . & . \\ . & . & . & . & . & . & . & . & . & . & . & . \\ 0 & 0 & 0 & 0 & 0 & . & -a_{WN-2} & a_{PN-2} & -a_{EN-2} & 0 \\ 0 & 0 & 0 & 0 & 0 & .0 & 0 & -a_{WN-1} & a_{PN-1} & -a_{EN-1} \\ 0 & 0 & 0 & 0 & 0 & .0 & 0 & 0 & -a_{WN} & a_{PN} \end{bmatrix}$$

$$X = \begin{bmatrix} \bar{c}_1 \\ \bar{c}_2 \\ \bar{c}_3 \\ \bar{c}_4 \\ . \\ . \\ \bar{c}_{N-2} \\ \bar{c}_{N-1} \\ \bar{c}_N \end{bmatrix} \quad and \quad B = \begin{bmatrix} b_1 \\ b_2 \\ b_3 \\ b_4 \\ . \\ . \\ b_{N-2} \\ b_{N-1} \\ b_N \end{bmatrix}$$

4.3.2 Solution of 2D Unsteady Concentration Equation

Using finite volume approach for solving the concentration partial differential equation, finite volume element is considered of volume ΔV within face point w and e at r direction and face point n and s at z direction. The considered finite volume is as in figure 4.5. Here, $\Delta V = 2\pi r dr dz$. Point P is the node where, concentration will be considered within the volume. w, e, n and s indicate the faces of the finite volume. W, E, N and S are the node point of the adjacent finite volume. Now, equation 4.7 is integrated twice within the finite ΔV and time limit t to $t + \Delta t$.

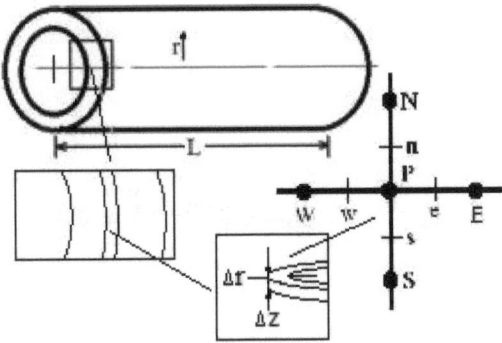

Fig.4.5: 2D finite volume in cylindrical system

$$\int_t^{t+\Delta t} \int_{\Delta V} \frac{\partial c}{\partial t} dV dt$$

$$= \int_t^{t+\Delta t} \int_{\Delta V} D \frac{1}{r} \frac{\partial}{\partial r}\left(r \frac{\partial c}{\partial r}\right) dV dt + + \int_t^{t+\Delta t} \int_{\Delta V} D \frac{\partial^2 c}{\partial z^2} dV dt$$

$$- \int_t^{t+\Delta t} \int_{\Delta V} u \frac{\partial c}{\partial r} dV dt$$

$$\Rightarrow \int_{\Delta V} \left[\int_t^{t+\Delta t} \frac{\partial c}{\partial t} dt\right] dV$$

$$= \int_t^{t+\Delta t} \left[\int_w^e D \frac{1}{r} \frac{\partial}{\partial r}\left(r \frac{\partial c}{\partial r}\right) 2\pi r dr dz\right] dt + \int_t^{t+\Delta t} \left[\int_s^n D \frac{\partial}{\partial z}\left(\frac{\partial c}{\partial z}\right) 2\pi r dr dz\right] dt$$

$$- \int_t^{t+\Delta t} \left[\int_w^e u \frac{\partial c}{\partial r} 2\pi r dr dz\right] dt$$

$$\Rightarrow \frac{c_P - c_P^0}{\Delta t} \Delta t \Delta V$$

$$= \int_t^{t+\Delta t} \left[2\pi D \Delta z \left(r \frac{\partial c}{\partial r}\right)\right]_w^e dt + \int_t^{t+\Delta t} \left[2\pi Dr \Delta r \left(\frac{\partial c}{\partial z}\right)\right]_s^n dt$$

$$- \int_t^{t+\Delta t} [2\pi u \Delta z r c]_w^e dt$$

$$\Rightarrow 2\pi r_P \Delta r \Delta z (c_P - c_P^0)$$

$$= \int_t^{t+\Delta t} 2\pi D \Delta z \left[\left(r \frac{\partial c}{\partial r}\right)_e - \left(r \frac{\partial c}{\partial r}\right)_w\right] + \int_t^{t+\Delta t} 2\pi D \Delta r \left[\left(r \frac{\partial c}{\partial z}\right)_n - \left(r \frac{\partial c}{\partial z}\right)_s\right]$$

$$- \int_t^{t+\Delta t} 2\pi u \Delta z [(rc)_e - (rc)_w]$$

Integrating time integral with θ time fraction, where θ fraction of previous time step and (1 − θ) time fraction of the current time step is taken into account.

$$2\pi r_P \Delta r \Delta z (c_P - c_P^0)$$

$$= \left[2\pi D \Delta z \left\{r_e \left(\frac{\partial c}{\partial r}\right)_e - r_w \left(\frac{\partial c}{\partial r}\right)_w\right\}\right]^\theta \Delta t$$

$$+ \left[2\pi D \Delta r \left\{r_n \left(\frac{\partial c}{\partial z}\right)_n - r_s \left(\frac{\partial c}{\partial z}\right)_s\right\}\right]^\theta \Delta t - [2\pi u \Delta z (r_e c_e - r_w c_w)]^\theta \Delta t$$

$$\Rightarrow \frac{2\pi r_P \Delta r \Delta z}{\Delta t}(c_P - c_P^0)$$

$$= 2\pi D \Delta z \theta r_e \frac{c_E - c_P}{\Delta r} - 2\pi D \Delta z \theta r_w \frac{c_P - c_W}{\Delta r} + 2\pi D \Delta z (1-\theta) r_e \frac{c_E^0 - c_P^0}{\Delta r}$$

$$- 2\pi D \Delta z (1-\theta) r_w \frac{c_P^0 - c_W^0}{\Delta r} + 2\pi D \Delta r \theta r_n \frac{c_N - c_P}{\Delta z} - 2\pi D \Delta r \theta r_s \frac{c_P - c_S}{\Delta z}$$

$$+ 2\pi D \Delta r (1-\theta) r_n \frac{c_N^0 - c_P^0}{\Delta z} - 2\pi D \Delta r (1-\theta) r_s \frac{c_P^0 - c_S^0}{\Delta z}$$

$$- 2\pi u \Delta z \theta r_e \frac{c_E + c_P}{2} + 2\pi u \Delta z \theta r_w \frac{c_W + c_P}{2} - 2\pi u \Delta z (1-\theta) r_e \frac{c_E^0 + c_P^0}{2}$$

$$+ 2\pi u \Delta z (1-\theta) r_w \frac{c_W^0 + c_P^0}{2} \qquad (4.36)$$

Now dividing both side with $2\pi D$, and setting $r_s = r_n = r_P$; as all the points S, N and P are at same radial position

$$\frac{r_P \Delta r \Delta z}{D \Delta t}(c_P - c_P^0)$$

$$= \frac{\Delta z \theta r_e}{\Delta r}(c_E - c_P) - \frac{\Delta z \theta r_w}{\Delta r}(c_P - c_W) + \frac{\Delta z (1-\theta) r_e}{\Delta r}(c_E^0 - c_P^0)$$

$$- \frac{\Delta z (1-\theta) r_w}{\Delta r}(c_P^0 - c_W^0) + \frac{\Delta r \theta r_P}{\Delta z}(c_N + c_S - 2c_P)$$

$$+ \frac{\Delta r (1-\theta) r_P}{\Delta z}(c_N^0 + c_S^0 - 2c_P^0) - \frac{u \Delta z \theta r_e}{2D}(c_E + c_P)$$

$$+ \frac{u \Delta z \theta r_w}{2D}(c_W + c_P) - \frac{u \Delta z (1-\theta) r_e}{2}(c_E^0 + c_P^0)$$

$$+ \frac{u \Delta z (1-\theta) r_w}{2}(c_W^0 + c_P^0)$$

$$\Rightarrow \left[\frac{r_P \Delta r \Delta z}{D \Delta t} + \frac{\Delta z \theta r_e}{\Delta r} + \frac{\Delta z \theta r_w}{\Delta r} + 2\frac{\Delta r \theta r_P}{\Delta z} - \frac{u \Delta z \theta}{2}(r_w - r_e)\right] c_P$$

$$= \left[\frac{\Delta z \theta r_e}{\Delta r} - \frac{u \Delta z \theta r_e}{2D}\right] c_E + \left[\frac{\Delta z \theta r_w}{\Delta r} + \frac{u \Delta z \theta r_w}{2D}\right] c_W + \frac{\Delta r \theta r_P}{\Delta z} c_N$$

$$+ \frac{\Delta r \theta r_P}{\Delta z} c_S$$

$$+ \left[\frac{r_P \Delta r \Delta z}{D \Delta t} - \frac{\Delta z(1-\theta)r_e}{\Delta r} - \frac{\Delta z(1-\theta)r_w}{\Delta r} - 2\frac{\Delta r(1-\theta)r_P}{\Delta z}\right.$$

$$\left. + \frac{u \Delta z(1-\theta)}{2}(r_w - r_e)\right] c_P^0 + \left[\frac{\Delta z(1-\theta)r_e}{\Delta r} - \frac{u \Delta z(1-\theta)r_e}{2}\right] c_E^0$$

$$+ \left[\frac{\Delta z(1-\theta)r_w}{\Delta r} + \frac{u \Delta z(1-\theta)r_w}{2}\right] c_W^0 + \frac{\Delta r(1-\theta)r_P}{\Delta z} c_N^0$$

$$+ \frac{\Delta r(1-\theta)r_P}{\Delta z} c_S^0 \tag{4.37}$$

Equation 4.37 is the algebraic form of the governing equation. For different time steps and for different radial positions, this equation can be solved. Now for getting better result, taking implicit method of problem solving, θ is set to 1 and thus equation 4.37 become, where $r_w - r_e = \Delta r$

$$\left[\frac{r_P \Delta r \Delta z}{D \Delta t} + \frac{\Delta z r_e}{\Delta r} + \frac{\Delta z r_w}{\Delta r} + 2\frac{\Delta r r_P}{\Delta z} - \frac{u \Delta z \Delta r}{2}\right] c_P$$

$$= \left[\frac{\Delta z r_e}{\Delta r} - \frac{u \Delta z r_e}{2D}\right] c_E + \left[\frac{\Delta z r_w}{\Delta r} + \frac{u \Delta z r_w}{2D}\right] c_W + \frac{\Delta r r_P}{\Delta z} c_N + \frac{\Delta r r_P}{\Delta z} c_S$$

$$+ \frac{r_P \Delta r \Delta z}{D \Delta t} c_P^0$$

$$\Rightarrow -\left[\frac{\Delta z r_w}{\Delta r} + \frac{u \Delta z r_w}{2D}\right] c_W + \left[\frac{r_P \Delta r \Delta z}{D \Delta t} + \frac{\Delta z r_e}{\Delta r} + \frac{\Delta z r_w}{\Delta r} + 2\frac{\Delta r r_P}{\Delta z} - \frac{u \Delta z \Delta r}{2}\right] c_P$$

$$- \left[\frac{\Delta z r_e}{\Delta r} - \frac{u \Delta z r_e}{2D}\right] c_E$$

$$= \frac{\Delta r r_P}{\Delta z} c_N + \frac{\Delta r r_P}{\Delta z} c_S + \frac{r_P \Delta r \Delta z}{D \Delta t} c_P^0 \tag{4.38}$$

Equation 4.38 is the final algebraic equation of the evaluating the concentration profile in unsteady 2 dimensional case of drug release. This equation can be solved by Tri-Diagonal Matrix Algorithm or by Gaussian algorithm by constructing a tri-diagonal coefficient matrix for the variable concentration point c_W, c_P and c_E where variable concentration point c_N and c_S are fixed for a single iteration step.

This equation can be compared with the following equation

$$-a_W c_W + a_P c_P - a_E c_E = a_N c_N + a_S c_S + a_P^0 c_P^0 \tag{4.39}$$

Comparing equation 4.35 and 4.36 it is found that

$$a_{Wi} = \Delta z \left(\frac{r_w}{\Delta r} + \frac{u r_w}{2D}\right) = \Delta z \left\{\left(\frac{1}{\Delta r} + \frac{u}{2D}\right)\left(\frac{r_P + r_W}{2}\right)\right\} = \frac{\Delta z}{2}\left\{\left(\frac{1}{\Delta r} + \frac{u}{2D}\right)(r_P + r_W)\right\}$$

$$a_{Ei} = \Delta z \left(\frac{r_e}{\Delta r} - \frac{u r_e}{2D}\right) = \Delta z \left\{\left(\frac{1}{\Delta r} - \frac{u}{2D}\right)\left(\frac{r_P + r_E}{2}\right)\right\} = \frac{\Delta z}{2}\left\{\left(\frac{1}{\Delta r} - \frac{u}{2D}\right)(r_P + r_E)\right\}$$

$$a_{Si} = a_{Ni} = \frac{\Delta r r_P}{\Delta z}$$

$$a_{Pi}^0 = \frac{r_P \Delta r \Delta z}{D \Delta t}$$

$$a_{Pi} = \frac{r_P \Delta r \Delta z}{D \Delta t} + \frac{\Delta z r_e}{\Delta r} + \frac{\Delta z r_w}{\Delta r} + 2\frac{\Delta r r_P}{\Delta z} - \frac{u \Delta z \Delta r}{2}$$

$$= \frac{r_P \Delta r \Delta z}{D \Delta t} + \Delta z \left(\frac{r_w}{\Delta r} - \frac{1}{2} + \frac{u r_w}{2D}\right) + \Delta z \left(\frac{r_e}{\Delta r} + \frac{1}{2} - \frac{u r_e}{2D}\right) + \frac{\Delta r r_P}{\Delta z} + \frac{\Delta r r_P}{\Delta z}$$

$$= a_{Pi}^0 + a_{Ei} + a_{Wi} + a_{Ni} + a_{Si}$$

$$b_i = a_N c_N + a_S c_S + a_P^0 c_P^0$$

Thus, to solve equation 4.38 according to equation 4.35, i = 2 to (N-1) the elements of coefficient matrix can be found from the value of a_{Pi}, a_{Ei} and a_{Wi}. The value of the column matrix can be found from b_i. Where, the variable matrix contain only the concentration terms of the radial direction.

To get the values of the first column of the coefficient matrix that means to get the inner wall concentration terms, equation 4.36 is modified in the following way by taking implicit scheme directly ($\theta = 1$)

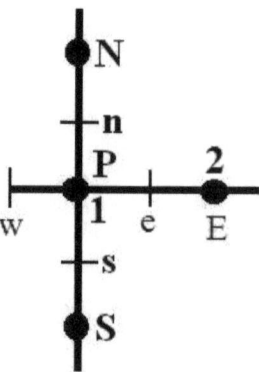

Fig. 4.6: Inner wall face boundary and node point of 2D discretized volume

$$\frac{2\pi r_P \Delta r \Delta z}{\Delta t}(c_P - c_P^0)$$

$$= 2\pi D \Delta z r_e \frac{c_E - c_P}{\Delta r} - 2\pi D \Delta z r_w \frac{c_P - c_w}{\frac{\Delta r}{2}} + 2\pi D \Delta r r_n \frac{c_N - c_P}{\Delta z}$$

$$- 2\pi D \Delta r r_s \frac{c_P - c_S}{\Delta z} - 2\pi u \Delta z r_e \frac{c_E + c_P}{2} + 2\pi u \Delta z r_w c_w$$

$$\Rightarrow \left[\frac{r_P \Delta r \Delta z}{D \Delta t} + \frac{\Delta z r_e}{\Delta r} + 2\frac{\Delta z r_w}{\Delta r} + 2\frac{\Delta r r_P}{\Delta z} + \frac{u \Delta z r_e}{2D}\right] c_P$$

$$= \left[\frac{\Delta z r_e}{\Delta r} - \frac{u \Delta z r_e}{2D}\right] c_E + \left[2\frac{\Delta z r_w}{\Delta r} + \frac{u \Delta z r_w}{D}\right] c_w + \frac{\Delta r r_P}{\Delta z} c_N + \frac{\Delta r r_P}{\Delta z} c_S$$

$$+ \frac{r_P \Delta r \Delta z}{D \Delta t} c_P^0$$

$$\Rightarrow \left[\frac{r_P \Delta r \Delta z}{D \Delta t} + \frac{\Delta z r_e}{\Delta r} + 2\frac{\Delta z r_w}{\Delta r} + 2\frac{\Delta r r_P}{\Delta z} + \frac{u \Delta z r_e}{2D}\right] c_P - \Delta z \left(\frac{r_e}{\Delta r} - \frac{u r_e}{2D}\right) c_E$$

$$= \left[2\frac{\Delta z r_w}{\Delta r} + \frac{u \Delta z r_w}{D}\right] c_w + \frac{\Delta r r_P}{\Delta z} c_N + \frac{\Delta r r_P}{\Delta z} c_S + \frac{r_P \Delta r \Delta z}{D \Delta t} c_P^0 \quad (4.40)$$

This equation can be compared with this equation

$$a_{P1} c_{P1} - a_{E1} c_{E1} = a_{w1} c_{w1} + a_{N1} c_{N1} + a_{S1} c_{S1} + a_{P1}^0 c_{P1}^0$$

Where,

$$a_{w1} = \Delta z \left(2 \frac{r_w}{\Delta r} + \frac{u r_w}{D}\right) = \Delta z \left\{r_w \left(\frac{2}{\Delta r} + \frac{u}{D}\right)\right\}$$

$$a_{E1} = \Delta z \left(\frac{r_e}{\Delta r} - \frac{ur_e}{2D}\right) = \Delta z \left\{\left(\frac{1}{\Delta r} - \frac{u}{2D}\right)\left(\frac{r_P + r_E}{2}\right)\right\} = \frac{\Delta z}{2}\left\{\left(\frac{1}{\Delta r} - \frac{u}{2D}\right)(r_P + r_E)\right\}$$

$$a_{S1} = a_{N1} = \frac{\Delta r r_P}{\Delta z}$$

$$a_{P1}^0 = \frac{r_P \Delta r \Delta z}{D \Delta t}$$

$$a_{P1} = \frac{r_P \Delta r \Delta z}{D \Delta t} + \frac{\Delta z r_e}{\Delta r} + 2\frac{\Delta z r_w}{\Delta r} + 2\frac{\Delta r r_P}{\Delta z} + \frac{u \Delta z r_e}{2D}$$

$$= \frac{r_P \Delta r \Delta z}{D \Delta t} + \Delta z \left(2\frac{r_w}{\Delta r} + \frac{ur_w}{D}\right) + \Delta z \left(\frac{r_e}{\Delta r} - \frac{ur_e}{2D}\right) + \frac{\Delta r r_P}{\Delta z} + \frac{\Delta r r_P}{\Delta z}$$

$$= a_{P1}^0 + a_{E1} + a_{w1} + a_{N1} + a_{S1}$$

$$b_1 = a_{w1} c_{w1} + a_{N1} c_{N1} + a_{S1} c_{S1} + a_{P1}^0 c_{P1}^0$$

To get the values of the last column of the coefficient matrix that means to get the outer wall concentration terms, equation 4.36 is modified in the following way by taking implicit scheme directly ($\theta = 1$)

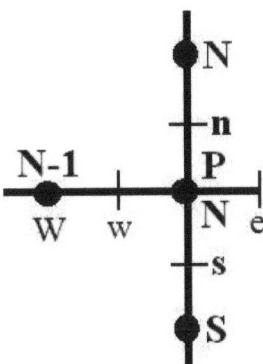

Fig. 4.7: Outer wall face boundary and node point of 2D discretized volume

$$\frac{2\pi r_P \Delta r \Delta z}{\Delta t}(c_P - c_P^0)$$

$$= 2\pi D \Delta z r_e \frac{c_e - c_P}{\Delta r/2} - 2\pi D \Delta z r_w \frac{c_P - c_W}{\Delta r} + 2\pi D \Delta r r_n \frac{c_N - c_P}{\Delta z}$$

$$- 2\pi D \Delta r r_s \frac{c_P - c_S}{\Delta z} - 2\pi u \Delta z r_e c_e + 2\pi u \Delta z r_w \frac{c_W + c_P}{2}$$

45

$$\Rightarrow \left[\frac{r_P \Delta r \Delta z}{D \Delta t} + 2\frac{\Delta z r_e}{\Delta r} + \frac{\Delta z r_w}{\Delta r} + 2\frac{\Delta r r_P}{\Delta z} - \frac{u \Delta z r_w}{2D}\right] c_P$$

$$= \left[2\frac{\Delta z r_e}{\Delta r} - \frac{u \Delta z r_e}{D}\right] c_e + \left[\frac{\Delta z r_w}{\Delta r} + \frac{u \Delta z r_w}{2D}\right] c_W + \frac{\Delta r r_P}{\Delta z} c_N + \frac{\Delta r r_P}{\Delta z} c_S$$

$$+ \frac{r_P \Delta r \Delta z}{D \Delta t} c_P^0$$

$$\Rightarrow -\Delta z \left(\frac{r_w}{\Delta r} + \frac{u r_w}{2D}\right) c_W + \left[\frac{r_P \Delta r \Delta z}{D \Delta t} + 2\frac{\Delta z r_e}{\Delta r} + \frac{\Delta z r_w}{\Delta r} + 2\frac{\Delta r r_P}{\Delta z} - \frac{u \Delta z r_w}{2D}\right] c_P$$

$$= \Delta z \left[2\frac{r_e}{\Delta r} - \frac{u r_e}{D}\right] c_e + \frac{\Delta r r_P}{\Delta z} c_N + \frac{\Delta r r_P}{\Delta z} c_S + \frac{r_P \Delta r \Delta z}{D \Delta t} c_P^0 \quad (4.41)$$

This equation can be compared with this equation

$$-a_{WN} c_{WN} + a_{PN} c_{PN} = a_{eN} c_{eN} + a_{NN} c_{NN} + a_{SN} c_{SN} + a_{PN}^0 c_{PN}^0$$

Where,

$$a_{eN} = \Delta z \left\{ r_e \left(\frac{2}{\Delta r} - \frac{u}{D}\right) \right\}$$

$$a_{WN} = \Delta z \left\{ r_w \left(\frac{1}{\Delta r} + \frac{u}{2D}\right) \right\} = \Delta z \left\{ \left(\frac{1}{\Delta r} + \frac{u}{2D}\right) \left(\frac{r_P + r_w}{2}\right) \right\}$$

$$= \frac{\Delta z}{2} \left\{ \left(\frac{1}{\Delta r} + \frac{u}{2D}\right) (r_P + r_w) \right\}$$

$$a_{SN} = a_{NN} = \frac{\Delta r r_P}{\Delta z}$$

$$a_{PN}^0 = \frac{r_P \Delta r \Delta z}{D \Delta t}$$

$$a_{PN} = \frac{r_P \Delta r \Delta z}{D \Delta t} + 2\frac{\Delta z r_e}{\Delta r} + \frac{\Delta z r_w}{\Delta r} + 2\frac{\Delta r r_P}{\Delta z} - \frac{u \Delta z r_w}{2D}$$

$$= \frac{r_P \Delta r \Delta z}{D \Delta t} + \Delta z \left\{ r_e \left(\frac{2}{\Delta r} - \frac{u}{D}\right) \right\} + \Delta z \left\{ r_w \left(\frac{1}{\Delta r} + \frac{u}{2D}\right) \right\} + \frac{\Delta r r_P}{\Delta z} + \frac{\Delta r r_P}{\Delta z}$$

$$= a_{PN}^0 + a_{eN} + a_{WN} + a_{NN} + a_{SN}$$

$$b_N = a_{eN} c_{eN} + a_{NN} c_{NN} + a_{SN} c_{SN} + a_{PN}^0 c_{PN}^0$$

4.3.3 Solution of Concentration Equation of BSMT

Using finite volume approach for solving the concentration partial differential equation, finite volume element is considered of volume ΔV within face point w and e at r direction and face point n and s at z direction. The considered finite volume is as in figure 4.5. Here, $\Delta V = 2\pi r dr dz$. Point P is the node where, concentration will be considered within the volume. w, e, n and s indicate the faces of the finite volume. W, E, N and S are the node point of the adjacent finite volume. Now, equation 4.9-i is integrated twice within the finite ΔV and time limit t to t + Δt.

$$\int_t^{t+\Delta t}\int_{\Delta V} \frac{\partial c_b}{\partial t} dVdt = \int_t^{t+\Delta t}\int_{\Delta V} D\frac{1}{r}\frac{\partial}{\partial r}\left(r\frac{\partial c_b}{\partial r}\right) dVdt - \int_t^{t+\Delta t}\int_{\Delta V} w\frac{\partial c_b}{\partial z} dVdt$$

$$\Rightarrow \int_{\Delta V}\left[\int_t^{t+\Delta t} \frac{\partial c_b}{\partial t} dt\right] dV$$

$$= \int_t^{t+\Delta t}\left[\int_w^e D\frac{1}{r}\frac{\partial}{\partial r}\left(r\frac{\partial c_b}{\partial r}\right) 2\pi r dr dz\right] dt - \int_t^{t+\Delta t}\left[\int_s^n w\frac{\partial c_b}{\partial z} 2\pi r dr dz\right] dt$$

$$\Rightarrow \frac{c_{bP} - c_{bP}^0}{\Delta t}\Delta t \Delta V = \int_t^{t+\Delta t}\left[2\pi D\Delta z\left(r\frac{\partial c_b}{\partial r}\right)\right]_w^e dt - \int_t^{t+\Delta t}[2\pi w \Delta r r c_b]_s^n dt$$

$$\Rightarrow 2\pi r_P \Delta r \Delta z(c_{bP} - c_{bP}^0)$$

$$= \int_t^{t+\Delta t} 2\pi D\Delta z\left[\left(r\frac{\partial c_b}{\partial r}\right)_e - \left(r\frac{\partial c_b}{\partial r}\right)_w\right] - \int_t^{t+\Delta t} 2\pi w \Delta r[(r c_b)_n - (r c_b)_s]$$

Integrating time integral with θ time fraction, where θ fraction of previous time step and (1 − θ) time fraction of the current time step is taken into account.

$2\pi r_P \Delta r \Delta z(c_{bP} - c_{bP}^0)$

$$= \left[2\pi D\Delta z\left\{r_e\left(\frac{\partial c_b}{\partial r}\right)_e - r_w\left(\frac{\partial c_b}{\partial r}\right)_w\right\}\right]^\theta \Delta t$$
$$- [2\pi w \Delta r(r_e c_{bn} - r_w c_{bs})]^\theta \Delta t$$

$$\Rightarrow \frac{2\pi r_p \Delta r \Delta z}{\Delta t}(c_{bP} - c_{bP}^0)$$

$$= 2\pi D \Delta z \theta r_e \frac{c_{bE} - c_{bP}}{\Delta r} - 2\pi D \Delta z \theta r_w \frac{c_{bP} - c_{bW}}{\Delta r}$$

$$+ 2\pi D \Delta z (1-\theta) r_e \frac{c_{bE}^0 - c_{bP}^0}{\Delta r} - 2\pi D \Delta z (1-\theta) r_w \frac{c_{bP}^0 - c_{bW}^0}{\Delta r}$$

$$- 2\pi w \Delta r \theta r_n \frac{c_{bN} + c_{bP}}{2} + 2\pi w \Delta r \theta r_s \frac{c_{bS} + c_{bP}}{2}$$

$$- 2\pi w \Delta r (1-\theta) r_n \frac{c_{bN}^0 + c_{bP}^0}{2}$$

$$+ 2\pi w \Delta r (1-\theta) r_s \frac{c_{bS}^0 + c_{bP}^0}{2} \tag{4.42}$$

Now dividing both side with 2πD, and setting $r_s = r_n = r_P$; as all the points S, N and P are at same radial position

$$\frac{r_p \Delta r \Delta z}{D \Delta t}(c_{bP} - c_{bP}^0)$$

$$= \frac{\Delta z \theta r_e}{\Delta r}(c_{bE} - c_{bP}) - \frac{\Delta z \theta r_w}{\Delta r}(c_{bP} - c_{bW}) + \frac{\Delta z (1-\theta) r_e}{\Delta r}(c_{bE}^0 - c_{bP}^0)$$

$$- \frac{\Delta z (1-\theta) r_w}{\Delta r}(c_{bP}^0 - c_{bW}^0) - \frac{w \Delta r \theta r_P}{2D}(c_{bN} - c_{bS})$$

$$- \frac{w \Delta r (1-\theta) r_P}{2}(c_{bN}^0 - c_{bS}^0)$$

$$\Rightarrow \left[\frac{r_p \Delta r \Delta z}{D \Delta t} + \frac{\Delta z \theta r_e}{\Delta r} + \frac{\Delta z \theta r_w}{\Delta r}\right] c_{bP}$$

$$= \left[\frac{\Delta z \theta r_e}{\Delta r}\right] c_{bE} + \left[\frac{\Delta z \theta r_w}{\Delta r}\right] c_{bW} - \frac{w \Delta r \theta r_P}{2D} c_{bN} + \frac{w \Delta r \theta r_P}{2D} c_{bS}$$

$$+ \left[\frac{r_p \Delta r \Delta z}{D \Delta t} - \frac{\Delta z (1-\theta) r_e}{\Delta r} - \frac{\Delta z (1-\theta) r_w}{\Delta r}\right] c_{bP}^0 + \left[\frac{\Delta z (1-\theta) r_e}{\Delta r}\right] c_{bE}^0$$

$$+ \left[\frac{\Delta z (1-\theta) r_w}{\Delta r}\right] c_{bW}^0 - \frac{w \Delta r (1-\theta) r_P}{2} c_{bN}^0$$

$$+ \frac{w \Delta r (1-\theta) r_P}{2} c_{bS}^0 \tag{4.43}$$

Equation 4.43 is the algebraic form of the governing equation. For different time steps and for different radial positions, this equation can be solved. Now for getting better result, taking implicit method of problem solving, θ is set to 1 and thus equation 4.43 become,

$$\left[\frac{r_P \Delta r \Delta z}{D \Delta t} + \frac{\Delta z r_e}{\Delta r} + \frac{\Delta z r_w}{\Delta r}\right] c_{bP}$$

$$= \left[\frac{\Delta z r_e}{\Delta r}\right] c_{bE} + \left[\frac{\Delta z r_w}{\Delta r}\right] c_{bW} - \frac{w \Delta r r_P}{2D} c_{bN} + \frac{w \Delta r r_P}{2D} c_{bS} + \frac{r_P \Delta r \Delta z}{D \Delta t} c_{bP}^0$$

$$\Rightarrow -\left[\frac{\Delta z r_w}{\Delta r}\right] c_{bW} + \left[\frac{r_P \Delta r \Delta z}{D \Delta t} + \frac{\Delta z r_e}{\Delta r} + \frac{\Delta z r_w}{\Delta r}\right] c_{bP} - \left[\frac{\Delta z r_e}{\Delta r}\right] c_{bE}$$

$$= -\frac{w \Delta r r_P}{2D} c_{bN} + \frac{w \Delta r r_P}{2D} c_{bS} + \frac{r_P \Delta r \Delta z}{D \Delta t} c_{bP}^0 \qquad (4.44)$$

Equation 4.44 is the final algebraic equation of the evaluating the concentration profile in unsteady BSMT drug release. This equation can be solved by Tri-Diagonal Matrix Algorithm or by Gaussian algorithm by constructing a tri-diagonal coefficient matrix for the variable concentration point c_{bW}, c_{bP} and c_{bE} where variable concentration point c_{bN} and c_{bS} are fixed for a single iteration step.

This equation can be compared with the following equation

$$-a_{bW} c_{bW} + a_{bP} c_{bP} - a_{bE} c_{bE} = a_{bN} c_{bN} + a_{bS} c_{bS} + a_{bP}^0 c_{bP}^0 \qquad (4.45)$$

Comparing equation 4.44 and 4.45 it is found that

$$a_{bWi} = \Delta z \left(\frac{r_w}{\Delta r}\right) = \Delta z \left\{\frac{1}{\Delta r}\left(\frac{r_P + r_W}{2}\right)\right\} = \frac{\Delta z}{2 \Delta r}(r_P + r_W)$$

$$a_{bEi} = \Delta z \left(\frac{r_e}{\Delta r}\right) = \Delta z \left\{\frac{1}{\Delta r}\left(\frac{r_P + r_E}{2}\right)\right\} = \frac{\Delta z}{2 \Delta r}(r_P + r_E)$$

$$a_{bSi} = -a_{bNi} = \frac{w \Delta r r_P}{2D}$$

$$a_{bPi}^0 = \frac{r_P \Delta r \Delta z}{D \Delta t}$$

$$a_{bPi} = \frac{r_P \Delta r \Delta z}{D \Delta t} + \frac{\Delta z r_e}{\Delta r} + \frac{\Delta z r_w}{\Delta r} = \frac{r_P \Delta r \Delta z}{D \Delta t} + \Delta z \left(\frac{r_w}{\Delta r}\right) + \Delta z \left(\frac{r_e}{\Delta r}\right) + \frac{w \Delta r r_P}{2D} - \frac{w \Delta r r_P}{2D}$$

$$= a_{bPi}^0 + a_{bEi} + a_{bWi} + a_{bNi} + a_{bSi}$$

$$b_i = a_{bN} c_{bN} + a_{bS} c_{bS} + a_{bP}^0 c_{bP}^0$$

Thus, to solve equation 4.44 according to equation 4.35, i = 2 to (N-1)th elements of coefficient matrix can be found from the value of a_{bPi}, a_{bEi} and a_{bWi}. The value of the column matrix can be found from b_i. Where, the variable matrix contain only the concentration terms of the radial direction.

To get the values of the first column of the coefficient matrix that means to get the inner wall concentration terms, equation 4.42 is modified in the following way by taking implicit scheme directly ($\theta = 1$)

$$\frac{2\pi r_P \Delta r \Delta z}{\Delta t}(c_{bP} - c_{bP}^0)$$

$$= 2\pi D \Delta z r_e \frac{c_{bE} - c_{bP}}{\Delta r} - 2\pi D \Delta z \theta r_w \frac{c_{bP} - c_{bw}}{\frac{\Delta r}{2}} - 2\pi w \Delta r r_n \frac{c_{bN} + c_{bP}}{2}$$

$$+ 2\pi w \Delta r r_s \frac{c_{bS} + c_{bP}}{2}$$

$$\Rightarrow \frac{r_P \Delta r \Delta z}{D \Delta t}(c_{bP} - c_{bP}^0)$$

$$= \frac{\Delta z r_e}{\Delta r}(c_{bE} - c_{bP}) - 2\frac{\Delta z r_w}{\Delta r}(c_{bP} - c_{bw}) + \frac{w \Delta r r_P}{2D}(c_{bN} - c_{bS})$$

$$\Rightarrow \left[\frac{r_P \Delta r \Delta z}{D \Delta t} + \frac{\Delta z r_e}{\Delta r} + 2\frac{\Delta z r_w}{\Delta r}\right]c_{bP}$$

$$= \left[\frac{\Delta z r_e}{\Delta r}\right]c_{bE} + \left[2\frac{\Delta z r_w}{\Delta r}\right]c_{bw} - \frac{w \Delta r r_P}{2D}c_{bN} + \frac{w \Delta r r_P}{2D}c_{bS} + \frac{r_P \Delta r \Delta z}{D \Delta t}c_{bP}^0$$

$$\Rightarrow \left[\frac{r_P \Delta r \Delta z}{D \Delta t} + \frac{\Delta z r_e}{\Delta r} + 2\frac{\Delta z r_w}{\Delta r}\right]c_{bP} - \left[\frac{\Delta z r_e}{\Delta r}\right]c_{bE}$$

$$= \left[2\frac{\Delta z r_w}{\Delta r}\right]c_{bw} - \frac{w \Delta r r_P}{2D}c_{bN} + \frac{w \Delta r r_P}{2D}c_{bS} + \frac{r_P \Delta r \Delta z}{D \Delta t}c_{bP}^0 \quad (4.46)$$

This equation can be compared with this equation

$$a_{bP1}c_{bP1} - a_{bE1}c_{bE1} = a_{bw1}c_{bw1} + a_{bN1}c_{bN1} + a_{bS1}c_{bS1} + a_{bP1}^0 c_{bP1}^0$$

Where,

$$a_{bw1} = \Delta z \left(2\frac{r_w}{\Delta r}\right) = \frac{\Delta z}{\Delta r}(r_w)$$

$$a_{bE1} = \Delta z \left(\frac{r_e}{\Delta r}\right) = \frac{\Delta z}{\Delta r}\left\{\left(\frac{r_P + r_E}{2}\right)\right\} = \frac{\Delta z}{2\Delta r}(r_P + r_E)$$

$$a_{bS1} = -a_{bN1} = \frac{w \Delta r r_P}{2D}$$

$$a_{bP1}^0 = \frac{r_P \Delta r \Delta z}{D \Delta t}$$

$$a_{bP1} = \frac{r_P \Delta r \Delta z}{D \Delta t} + \frac{\Delta z r_e}{\Delta r} + 2\frac{\Delta z r_w}{\Delta r} = \Delta z \left(2\frac{r_w}{\Delta r}\right) + \Delta z \left(\frac{r_e}{\Delta r}\right) + \frac{w \Delta r r_P}{2D} - \frac{w \Delta r r_P}{2D} + \frac{r_P \Delta r \Delta z}{D \Delta t}$$

$$= a_{bP1}^0 + a_{bE1} + a_{bw1} + a_{bN1} + a_{bS1}$$

$$b_1 = a_{bw1} c_{bw1} + a_{bN1} c_{bN1} + a_{bS1} c_{bS1} + a_{bP1}^0 c_{bP1}^0$$

To get the values of the last column of the coefficient matrix that means to get the outer wall concentration terms, equation 4.42 is modified in the following way by taking implicit scheme directly ($\theta = 1$)

$$\frac{2\pi r_P \Delta r \Delta z}{\Delta t}(c_{bP} - c_{bP}^0)$$

$$= 2\pi D \Delta z r_e \frac{c_{be} - c_{bP}}{\frac{\Delta r}{2}} - 2\pi D \Delta z \theta r_w \frac{c_{bP} - c_{bW}}{\Delta r} - 2\pi w \Delta r r_n \frac{c_{bN} + c_{bP}}{2}$$

$$+ 2\pi w \Delta r r_s \frac{c_{bS} + c_{bP}}{2}$$

$$\Rightarrow \frac{r_P \Delta r \Delta z}{D \Delta t}(c_{bP} - c_{bP}^0) = 2\Delta z r_e \frac{c_{be} - c_{bP}}{\Delta r} - \Delta z \theta r_w \frac{c_{bP} - c_{bW}}{\Delta r} - \frac{w \Delta r r_n}{2D}(c_{bN} - c_{bS})$$

$$\Rightarrow \left[\frac{r_P \Delta r \Delta z}{D \Delta t} + 2\frac{\Delta z r_e}{\Delta r} + \frac{\Delta z r_w}{\Delta r}\right] c_{bP}$$

$$= \left[2\frac{\Delta z r_e}{\Delta r}\right] c_{be} + \left[\frac{\Delta z r_w}{\Delta r}\right] c_{bW} - \frac{w \Delta r r_P}{2D} c_{bN} + \frac{w \Delta r r_P}{2D} c_{bS} + \frac{r_P \Delta r \Delta z}{D \Delta t} c_{bP}^0$$

$$\Rightarrow -\left[\frac{\Delta z r_w}{\Delta r}\right] c_{bW} + \left[\frac{r_P \Delta r \Delta z}{D \Delta t} + 2\frac{\Delta z r_e}{\Delta r} + \frac{\Delta z r_w}{\Delta r}\right] c_{bP}$$

$$= \left[2\frac{\Delta z r_e}{\Delta r}\right] c_{be} - \frac{w \Delta r r_P}{2D} c_{bN} + \frac{w \Delta r r_P}{2D} c_{bS} + \frac{r_P \Delta r \Delta z}{D \Delta t} c_{bP}^0 \quad (4.47)$$

This equation can be compared with this equation

$$-a_{bWN} c_{bWN} + a_{bPN} c_{bPN} = a_{beN} c_{beN} + a_{bNN} c_{bNN} + a_{bSN} c_{bSN} + a_{bPN}^0 c_{bPN}^0$$

Where,

$$a_{beN} = \Delta z \left(2\frac{r_e}{\Delta r}\right) = \frac{\Delta z}{\Delta r}(r_e)$$

$$a_{bWN} = \Delta z \left(\frac{r_w}{\Delta r}\right) = \Delta z \left\{\left(\frac{r_P + r_W}{2}\right)\right\} = \frac{\Delta z}{2\Delta r}(r_P + r_W)$$

$$a_{bSN} = -a_{bNN} = \frac{w\Delta r r_P}{2D}$$

$$a_{bPN}^0 = \frac{r_P \Delta r \Delta z}{D\Delta t}$$

$$a_{bPN} = \frac{r_P \Delta r \Delta z}{D\Delta t} + 2\frac{\Delta z r_e}{\Delta r} + \frac{\Delta z r_w}{\Delta r} = \frac{r_P \Delta r \Delta z}{D\Delta t} + \Delta z \left(\frac{r_w}{\Delta r}\right) + \Delta z \left(2\frac{r_e}{\Delta r}\right) + \frac{w\Delta r r_P}{2D} - \frac{w\Delta r r_P}{2D}$$

$$= a_{bPN}^0 + a_{bEN} + a_{bwN} + a_{bNN} + a_{bSN}$$

$$b_N = a_{beN} c_{beN} + a_{bNN} c_{bNN} + a_{bSN} c_{bSN} + a_{bPN}^0 c_{bPN}^0$$

4.3.4 Solution of Equation Describing Coating Thickness

From equation 4.8 can be solved analytically by the following way

$$\frac{\partial h}{\partial t} = -kh^n$$

$$\Rightarrow \frac{dh}{h^n} = -kdt$$

$$\Rightarrow \int_{h_0}^{h} \frac{dh}{h^n} = -\int_0^t kdt$$

$$\Rightarrow \left[\frac{h^{-n+1}}{-n+1}\right]_{h_0}^{h} = -k[t]_0^t$$

$$\Rightarrow \frac{1}{1-n}[h^{1-n} - h_0^{1-n}] = -kt$$

$$\Rightarrow h^{1-n} = h_0^{1-n} - kt(1-n)$$

$$\Rightarrow h = [h_0^{1-n} - kt(1-n)]^{\frac{1}{1-n}} \quad (4.48)$$

Using equation 4.48 the remaining thickness of the biodegradable drug coating in DES can be calculated, as no diffusion of drug has been assumed in the coating over DES.

Chapter Five

RESULTS AND DISCUSSION

In this chapter the simulation performed in the thesis using MATLAB has been represented graphically with the MATLAB codes. Both two governing equations and coating thickness equation has been solved using finite volume technique.

5.1 Solution of the Coating Thickness Equation

To solve equation 4.8, the analytical procedure has been shown in section 4.3.4 of Chapter Four. In this solution, there are two unknown parameter k and n which are dependent on coating material property of DES. The value of n and k has been optimized by the following way. At first the optimization has been performed setting n values less than one and varying k value within 1×10^{-10} to 3×10^{-10}. The desired value of n and k has been set by plotting 3D plot of exponent n vs. coating thickness vs. time plot. In this optimization it has been assumed that the initial coating thickness is $h_0 = 12.6 \times 10^{-6}$ m and coating thickness will last for 30 days or 43200 minutes time. MATLAB coding of this solution has been given in Appendix B.1

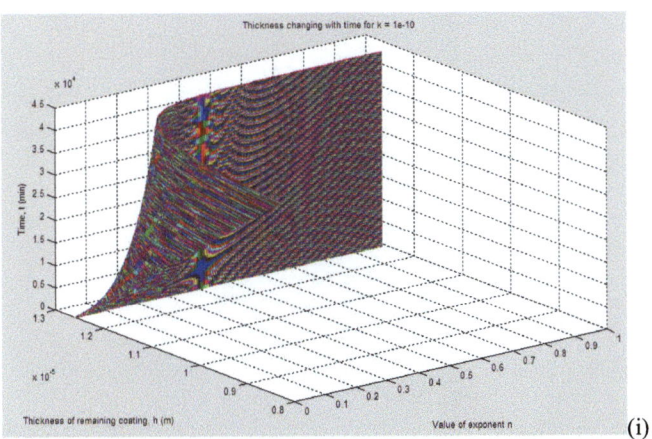

(i)

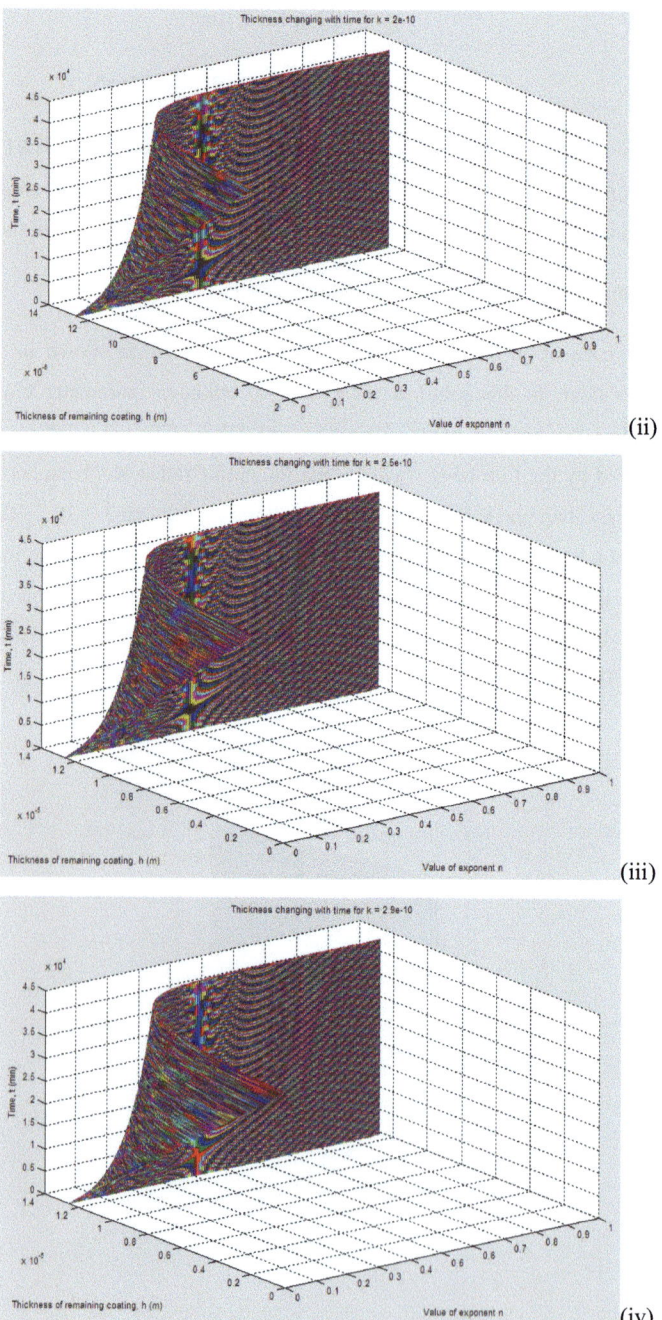

(ii)

(iii)

(iv)

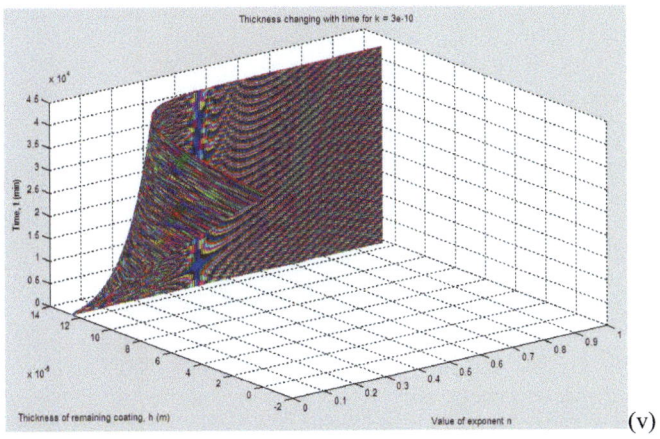 (v)

Fig. 5.1: Exponent n versus coating thickness versus time plot for (i) k = 1 × 10^{-10}, (ii) k = 2 × 10^{-10}, (iii) k = 2.5 × 10^{-10}, (iv) k = 2.9 × 10^{-10} and (v) k = 3 × 10^{-10}.

To understand these 3D thickness reduction plots properly contour plots are drawn where values of a certain variable is represented as a height above the plane of independent variables. The contour plots of above data are as follows

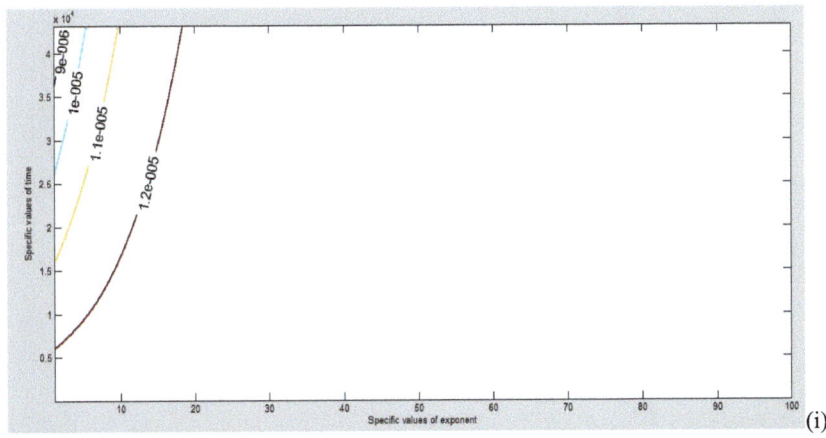 (i)

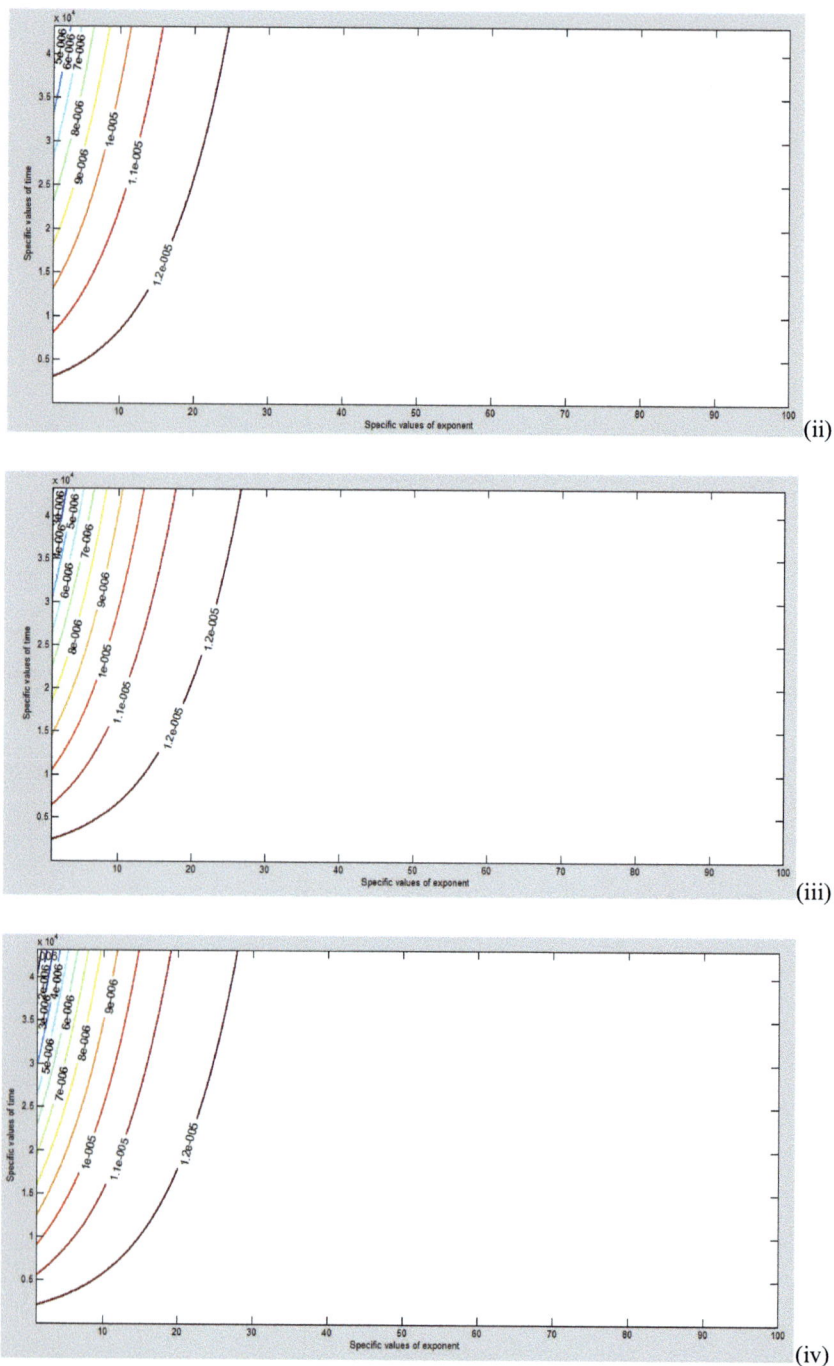

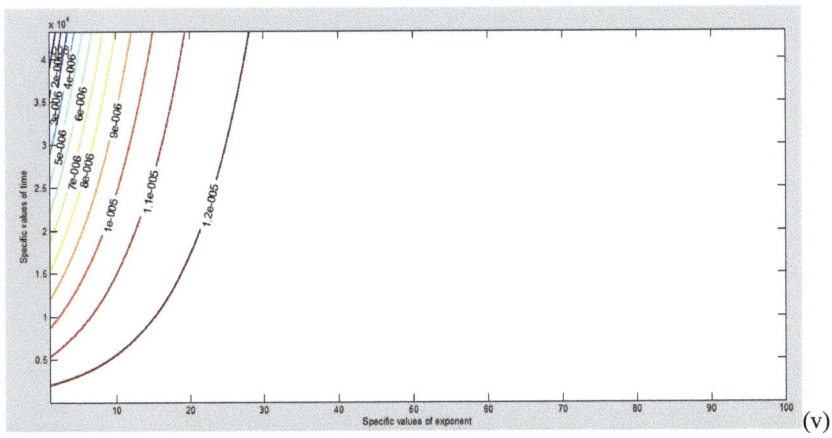
(v)

Fig. 5.2: Contour plot of the coating thickness at time versus radius plane for n values less than 1 and (i) k = 1 × 10^{-10}, (ii) k = 2 × 10^{-10}, (iii) k = 2.5 × 10^{-10}, (iv) k = 2.9 × 10^{-10} and (v) k = 3 × 10^{-10}. Coating thickness values mentioned in contour are in meter unit.

From contour plots of figure 5.2 it is found that for different values of k and range of exponent n, different lasting conditions of coating thickness are found. From all the observations it is found that at n = 0.0 and k = 2.9 × 10^{-10} coating thickness reduces to zero and for n = 0.01 and k = 3 × 10^{-10} coating thickness become negative.

It has been observed that, for k greater than 2.9 × 10^{-10} the coating thickness become negative which is not possible. The consequence for k > 2.9 × 10^{-10} can be shown in following figure 5.3. For n values less than 1, the best coating decay condition is found for n = 0 and k = 2.9 × 10^{-10}. This coating decay condition has been shown in figure 5.4 in coating thickness, h versus time plot.

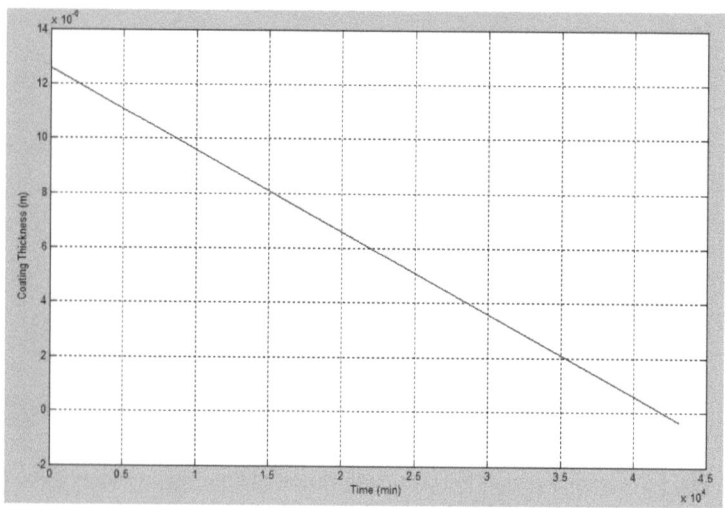

Fig. 5.3: Coating thickness versus time plot for n = 0 and k = 3 × 10^{-10}.

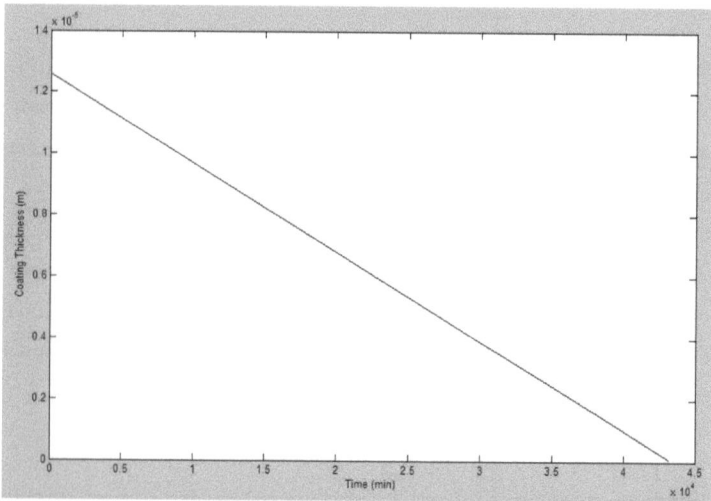

Fig. 5.4: Coating thickness versus time plot for n = 0 and k = 2.9 × 10^{-10}.

From figure 5.4 it is seen that DES coating thickness decreases linearly with time. Thus to get a better coating thickness decay condition, value of n and k are continued to optimize at range of n between 1 and 2 and range of k between 1 × 10^{-10} to 1.0. The 3D plots of n vs. h vs. time plots of this optimization are as follows.

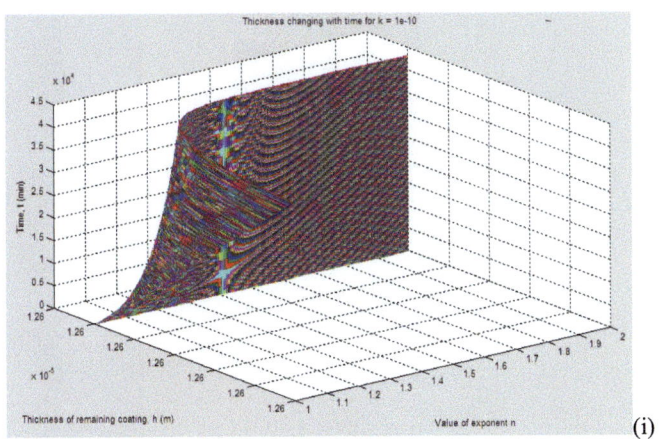 (i)

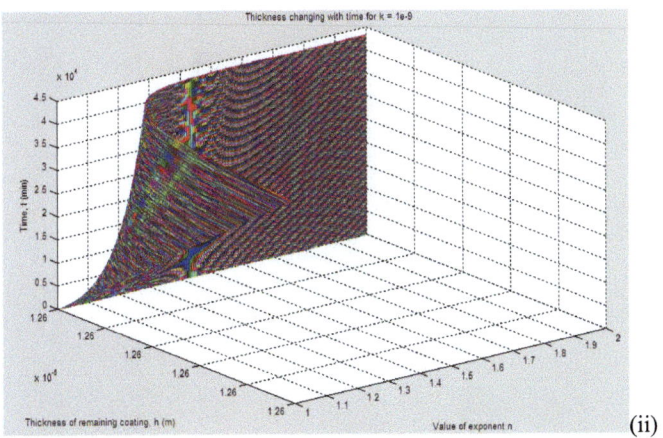 (ii)

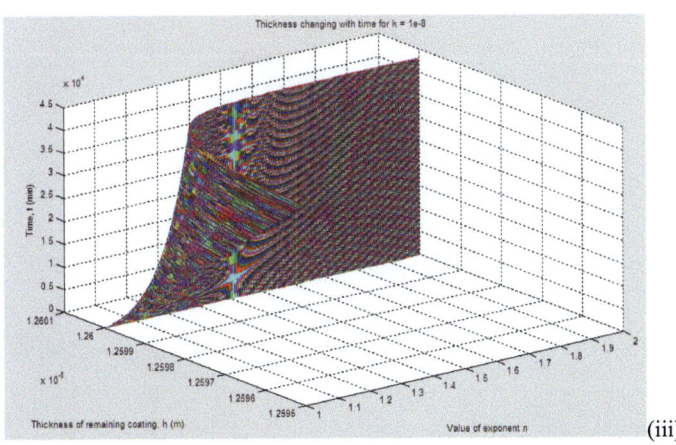 (iii)

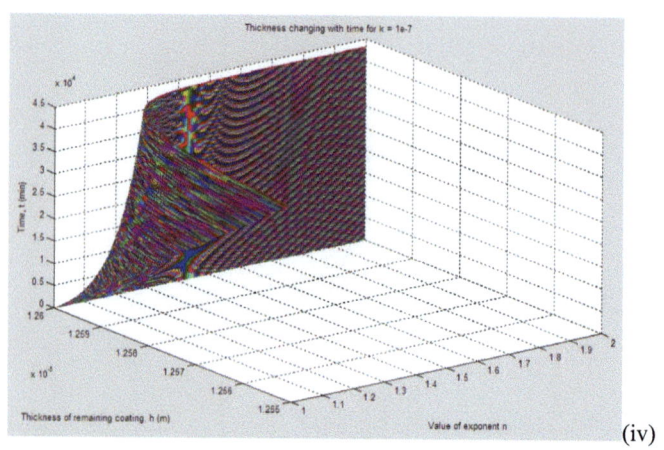

(iv)

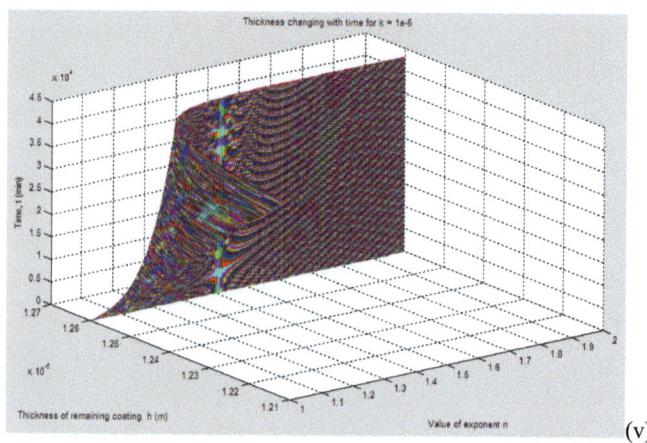

(v)

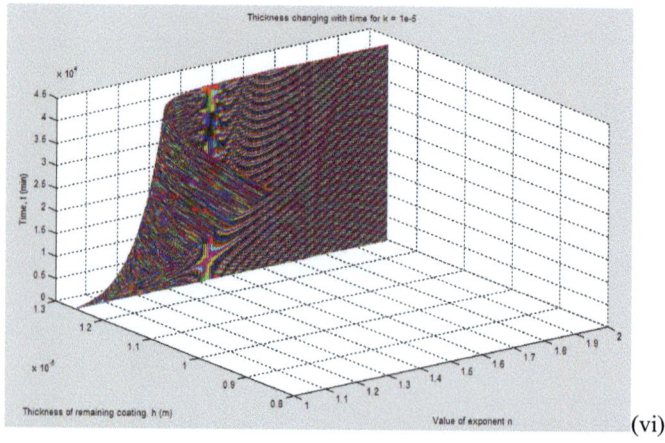

(vi)

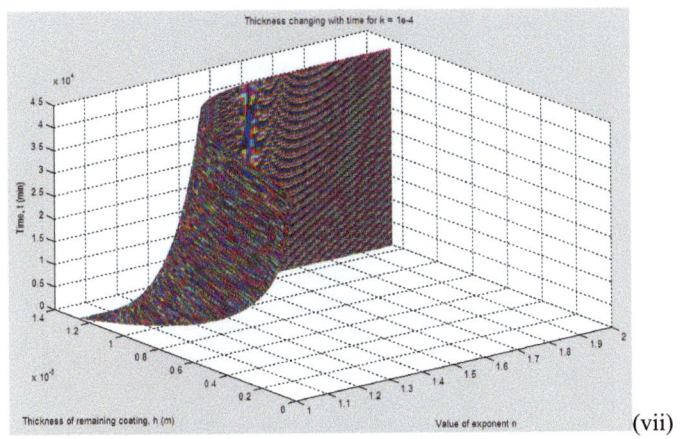

(vii)

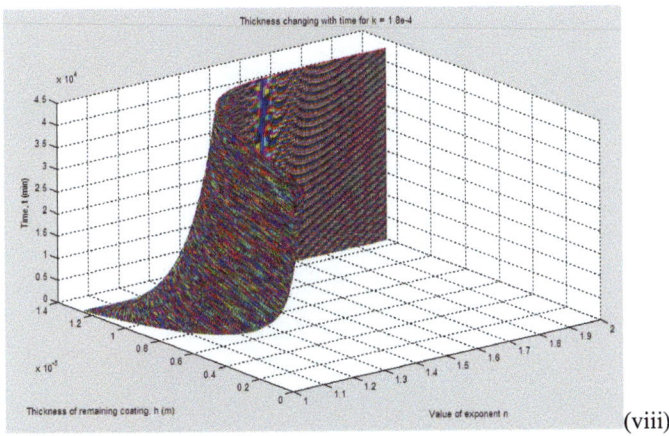

(viii)

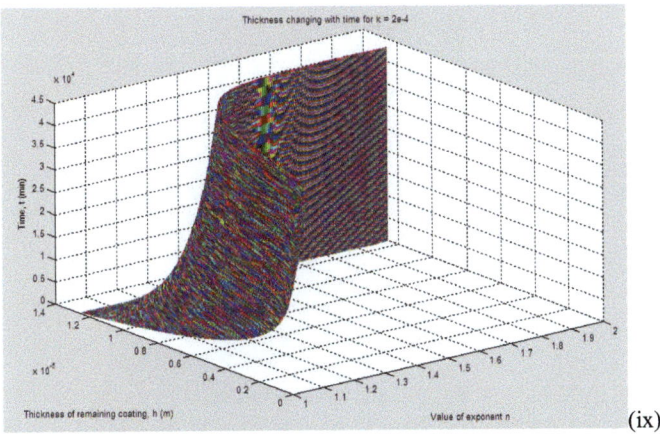

(ix)

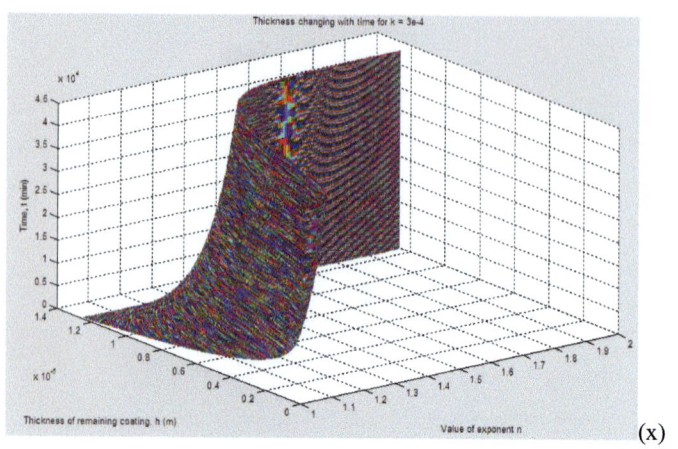
(x)

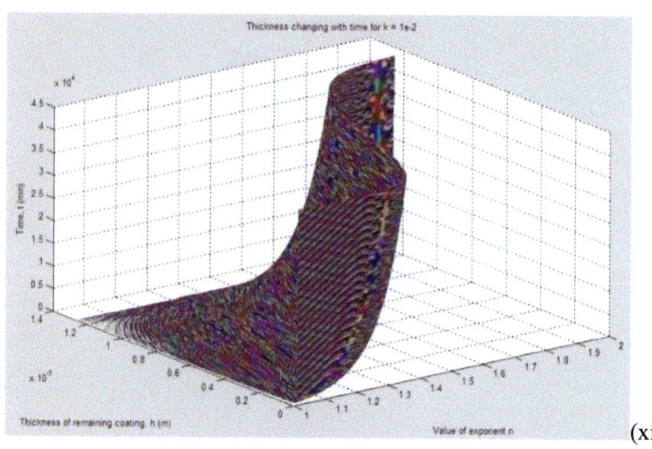
(xi)

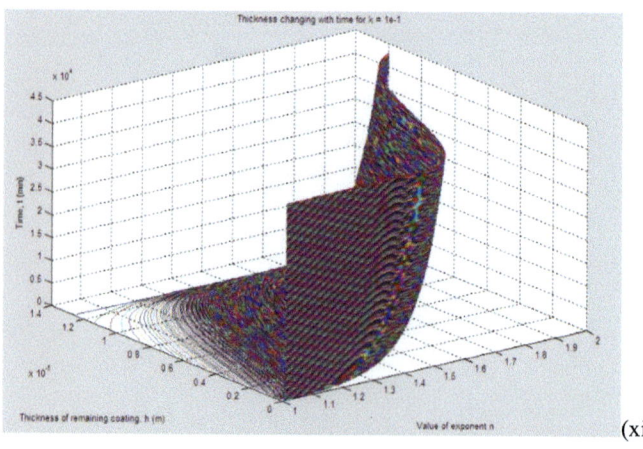
(xii)

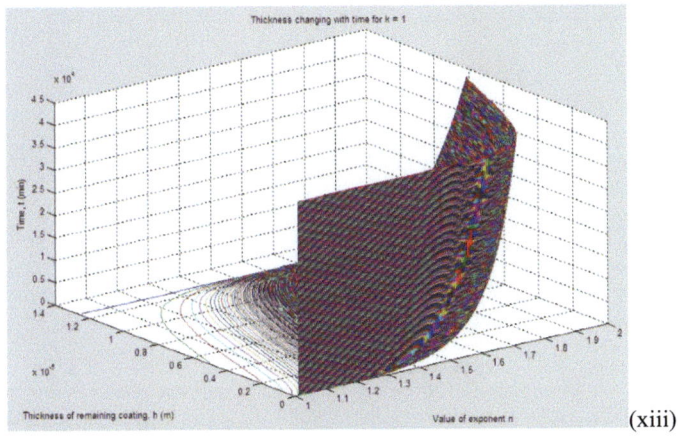
(xiii)

Fig. 5.5: Exponent n versus coating thickness versus time plot for (i) $k = 1 \times 10^{-10}$, (ii) $k = 1 \times 10^{-9}$, (iii) $k = 1 \times 10^{-8}$, (iv) $k = 1 \times 10^{-7}$, (v) $k = 1 \times 10^{-6}$, (vi) $k = 1 \times 10^{-5}$, (vii) $k = 1 \times 10^{-4}$, (viii) $k = 1.8 \times 10^{-4}$, (ix) $k = 2 \times 10^{-4}$, (x) $k = 3 \times 10^{-4}$, (xi) $k = 1 \times 10^{-2}$, (xii) $k = 1 \times 10^{-1}$ and (xiii) $k = 1.0$

To understand these 3D thickness reduction plots properly contour plots are drawn where values of a certain variable is represented as a height above the plane of independent variables. The contour plots of above data are as follows

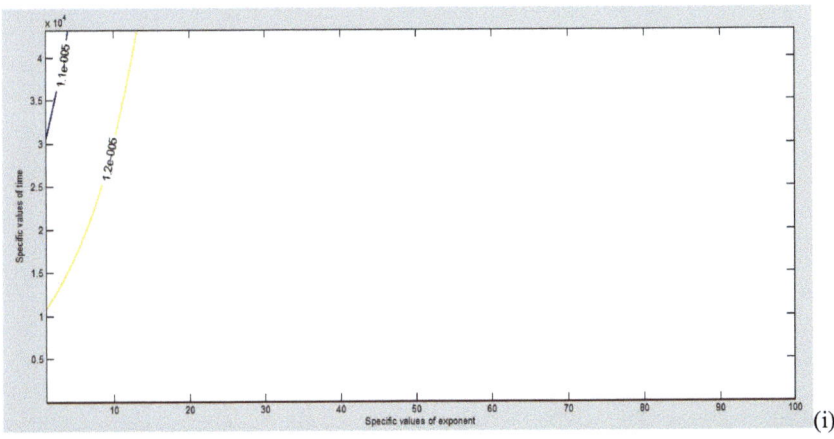
(i)

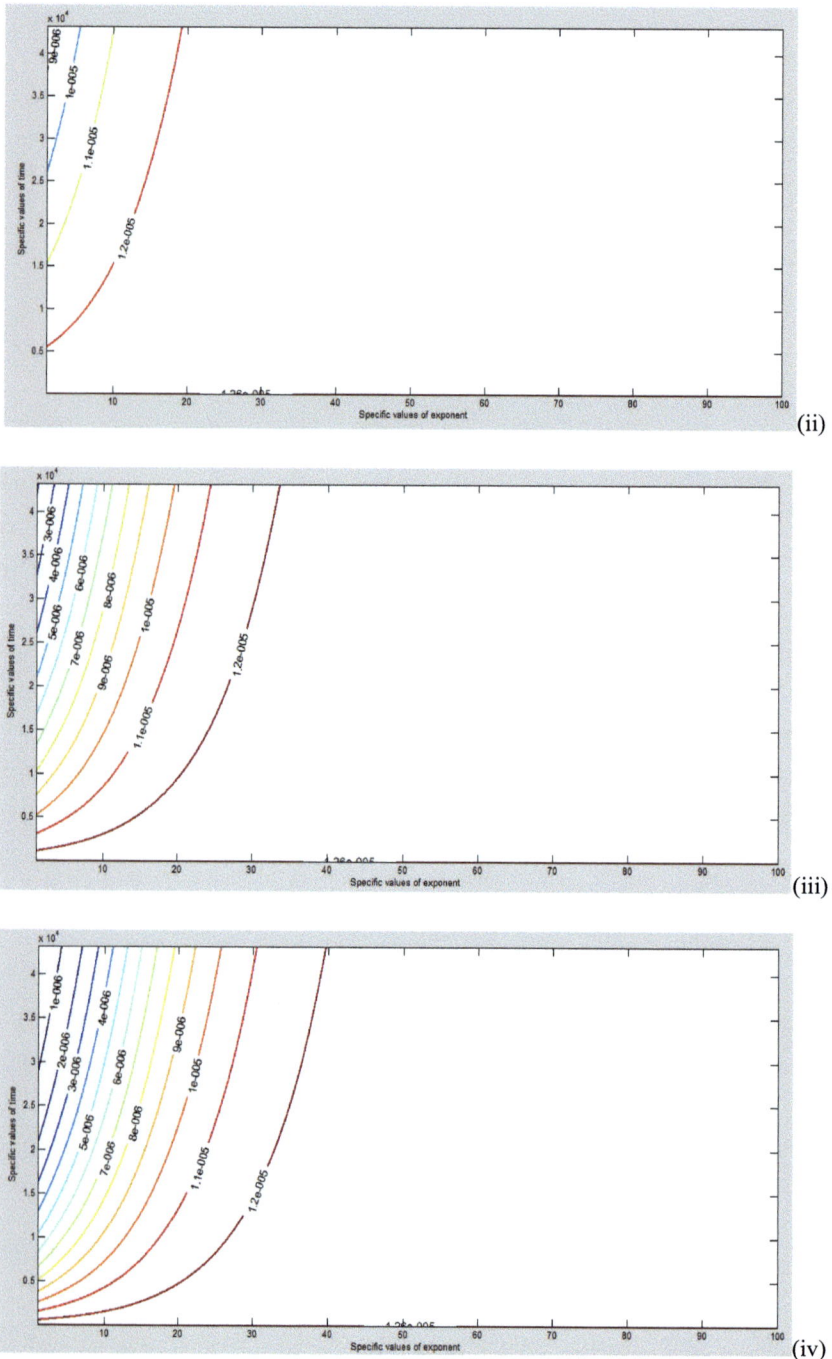

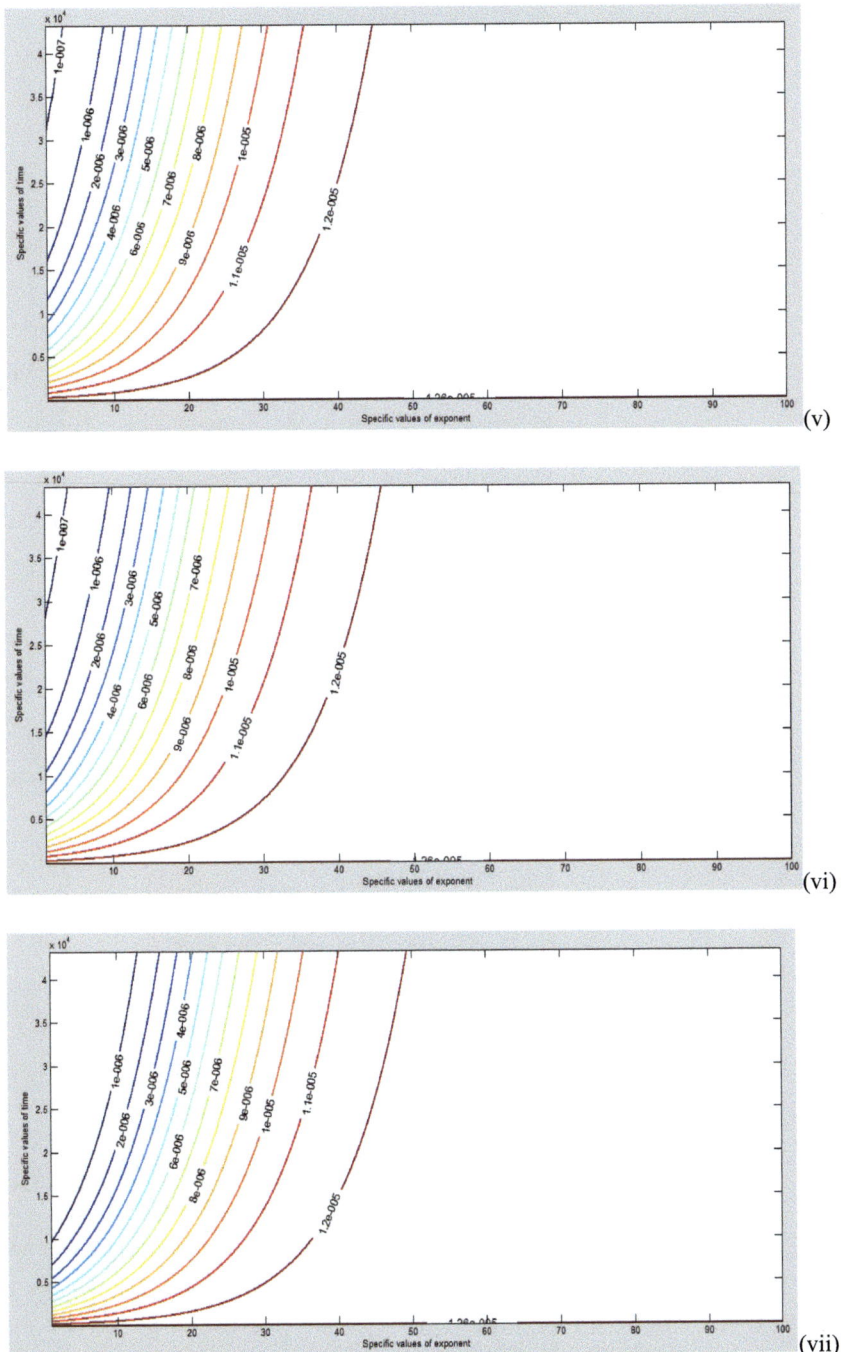

(v)

(vi)

(vii)

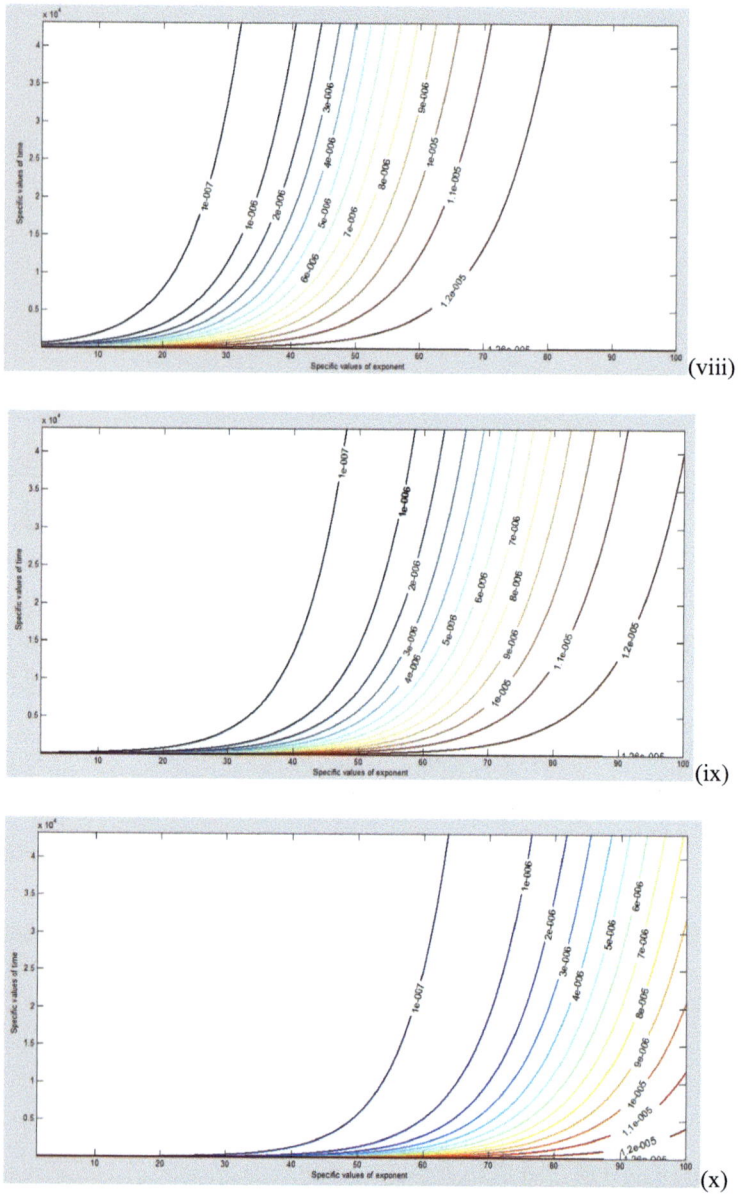

Fig. 5.6: Contour plot of the coating thickness at time versus radius plane for n values greater than 1 and (i) k = 5 × 10⁻⁶, (ii) k = 1 × 10⁻⁵, (iii) k = 5 × 10⁻⁵, (iv) k = 1 × 10⁻⁴, (v) k = 1.8 × 10⁻⁴, (vi) k = 2 × 10⁻⁴, (vii) k = 3 × 10⁻⁴, (viii) k = 1 × 10⁻², (ix) k = 1 × 10⁻¹ and (x) k = 1.0. Coating thickness values mentioned in contour are in meter unit.

From the above 3D and contour plot, it has been observed that with increasing value k the decay of the coating has been increased for increasing value of n also. This has been clearly understood from the contour plot. For k value from 1×10^{-10} to 1×10^{-6} almost no decay of the coating thickness is found. For $k = 5 \times 10^{-6}$ coating thickness decay till 11×10^{-6} m, for $k = 1 \times 10^{-5}$ thickness decay till 9×10^{-6} m and for $k = 5 \times 10^{-5}$ the remaining thickness value is 3×10^{-6} m. From all the other contour plot of figure 5.6, it is found that, at $k = 1.8 \times 10^{-6}$ and $n = 1.01$ best coating thickness decay condition is found, where thickness decay till 3×10^{-8} m. Thus this value of k and n is taken as optimum for the rest of the DES calculation. In case of k value greater than 1.8×10^{-6} other n values are also found where coating thickness tends to zero due to decay, but those decay faster than desired.

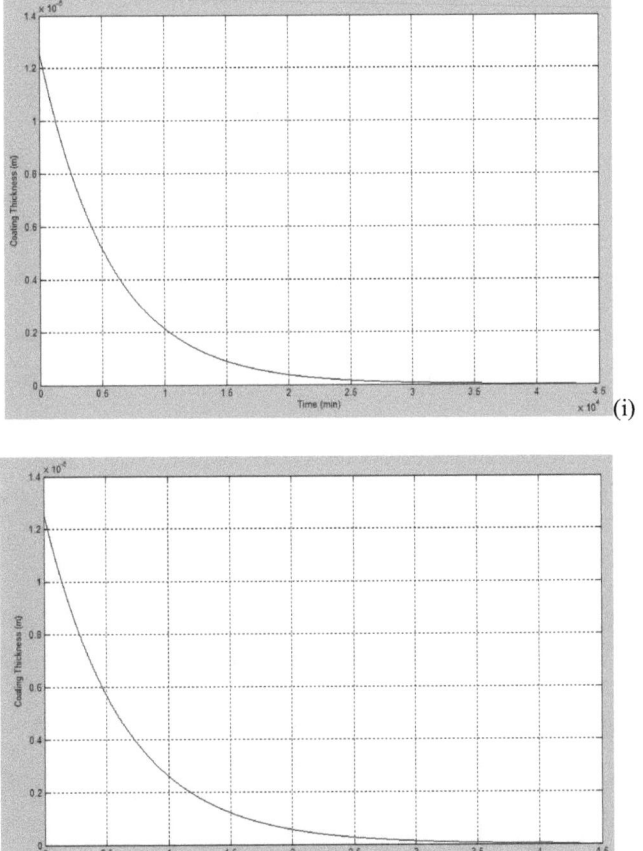

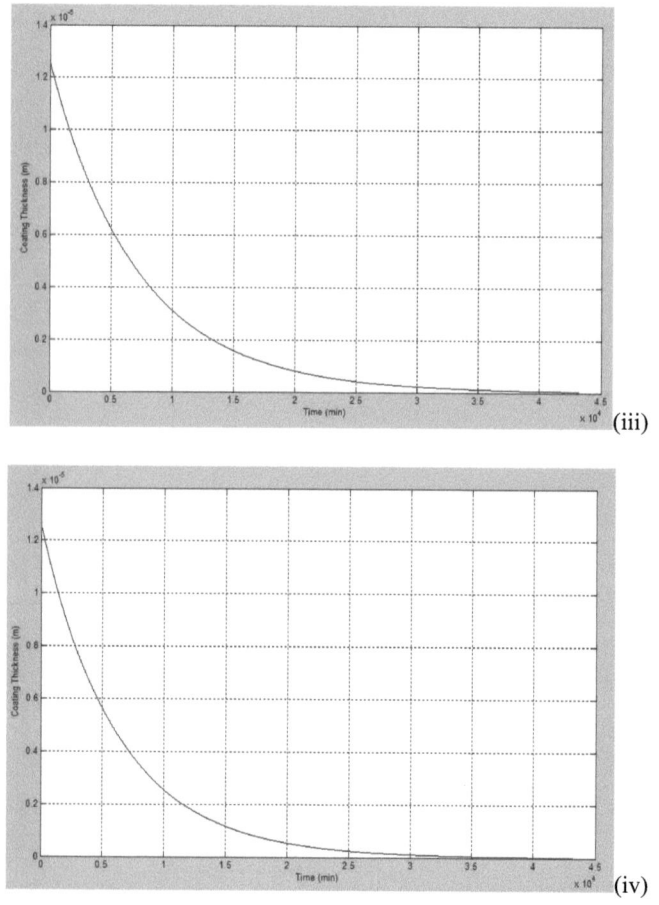

Fig. 5.7: DES coating thickness versus time plot for (i) $n = 1.01$, $k = 2 \times 10^{-4}$; (ii) $n = 1.02$, $k = 2 \times 10^{-4}$; (iii) $n = 1.03$, $k = 2 \times 10^{-4}$ and (iv) $n = 1.01$, $k = 1.8 \times 10^{-4}$.

From the coating thickness versus time plot of figure 5.7, it has been clear that $n = 1.01$, and $k = 1.8 \times 10^{-4}$ gives best decay condition DES coating thickness with time.

5.2 Solution of Unsteady 1D Concentration Governing Equation

The unsteady concentration equation in radial direction to get the drug concentration released from the DES coating at the artery wall was equation 4.6. This equation was solved using finite volume algorithm where three algebraic equation of concentration:

4.21, 4.27, 4.31 was found. Using these three equations, a tri-diagonal matrix was formed which was then solved using following MATLAB coding. In this solution procedure, DES coating thickness was taken as 12.6×10^{-6} m, Drug loading in the coating: 100 µg/cm^2 or 1 g/m^2, radius of the artery wall was varied between 1.5×10^{-3} m to 2.5×10^{-3} m, DES length was taken to be 8×10^{-3} m. The equation was solved for a time of 30 days or 43200 minutes, where discretized radius was 0.02×10^{-3} m and discretized time was 1 minute. Peclet number was taken as 1 in this solution. From the MATLAB coding of solution equation mentioned above, a 3D concentration profile was found with respect radial position and time which is as follows. MATLAB coding of this solution has been given in Appendix B.2

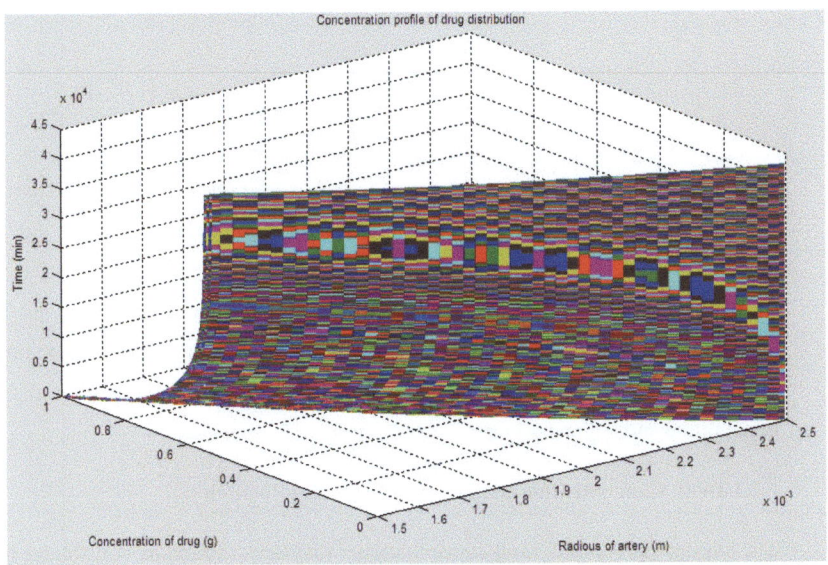

Fig. 5.8: 3D plot of radius versus drug concentration in artery wall versus time

For better understanding, this concentration profile can be represented in contour plot at the plane of relative time and radial position. In concentration contour plot of figure 5.9, it is clearly seen that, as radial position shift from inner artery wall (relative radial position 1) to outer artery wall (relative radial position 50) iso-concentration curve of DES drug release with time become flatter. These curves indicate that at inner artery wall drug concentration change is very rapid where at the outer artery wall it is sluggish.

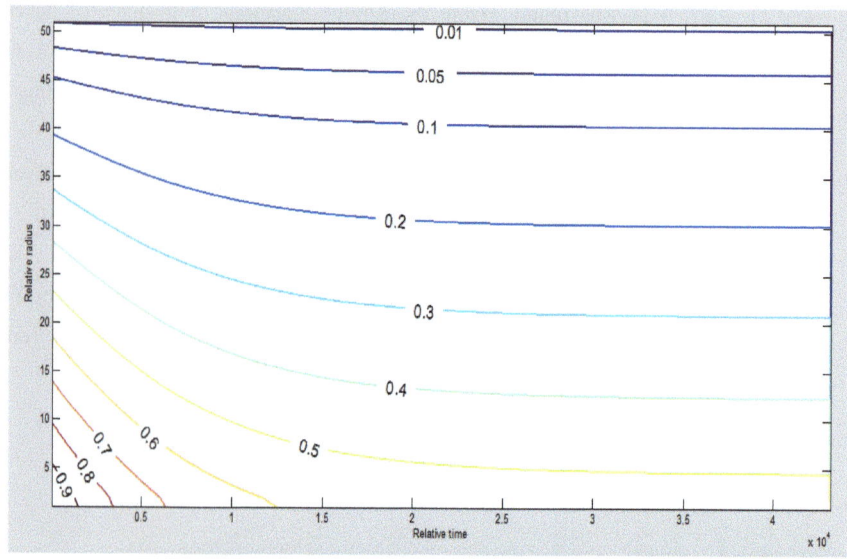

Fig. 5.9: Contour plot of concentration in relative radial and time plane. Drug concentration values mentioned in the contour are in g unit.

The concentration of drug in artery wall can analyzed from this concentration profile by taking drug concentration profile with respect to different radial position at different certain time and with respect to time at different radial position.

5.2.1 Drug Concentration with respect to Radial Position

The curves representing drug concentration with respect to radius at different time are as follows.

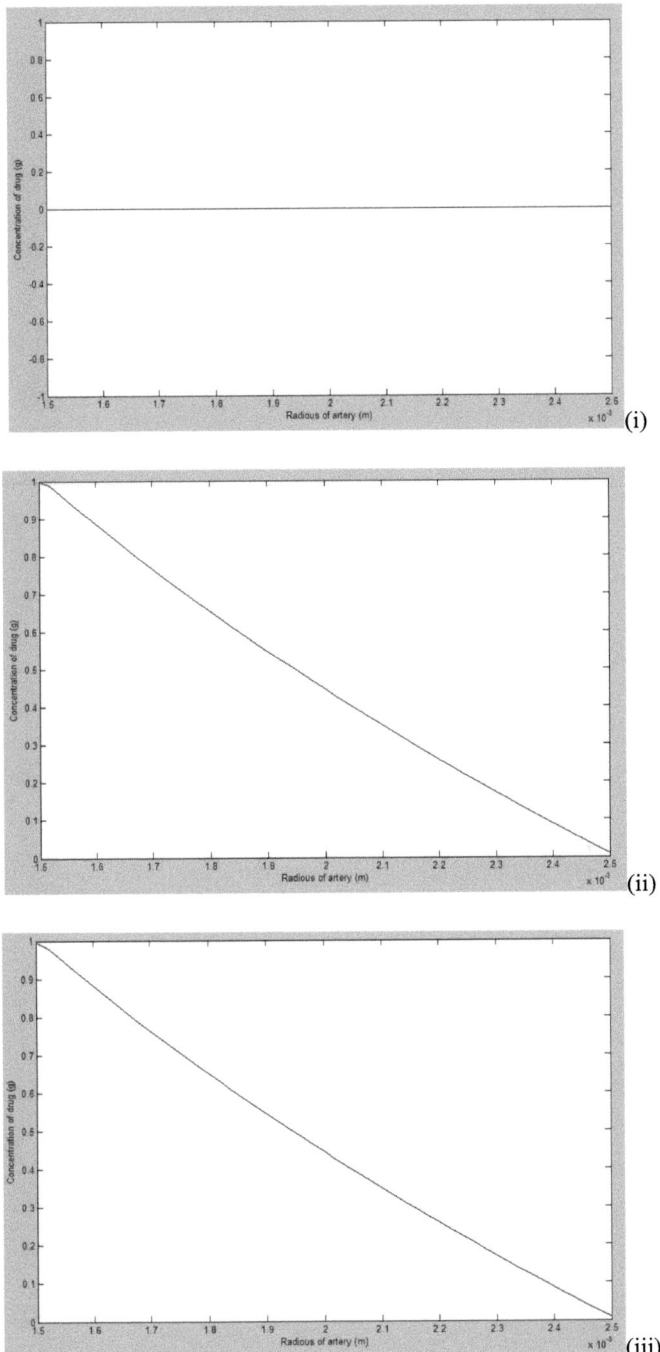

(i)

(ii)

(iii)

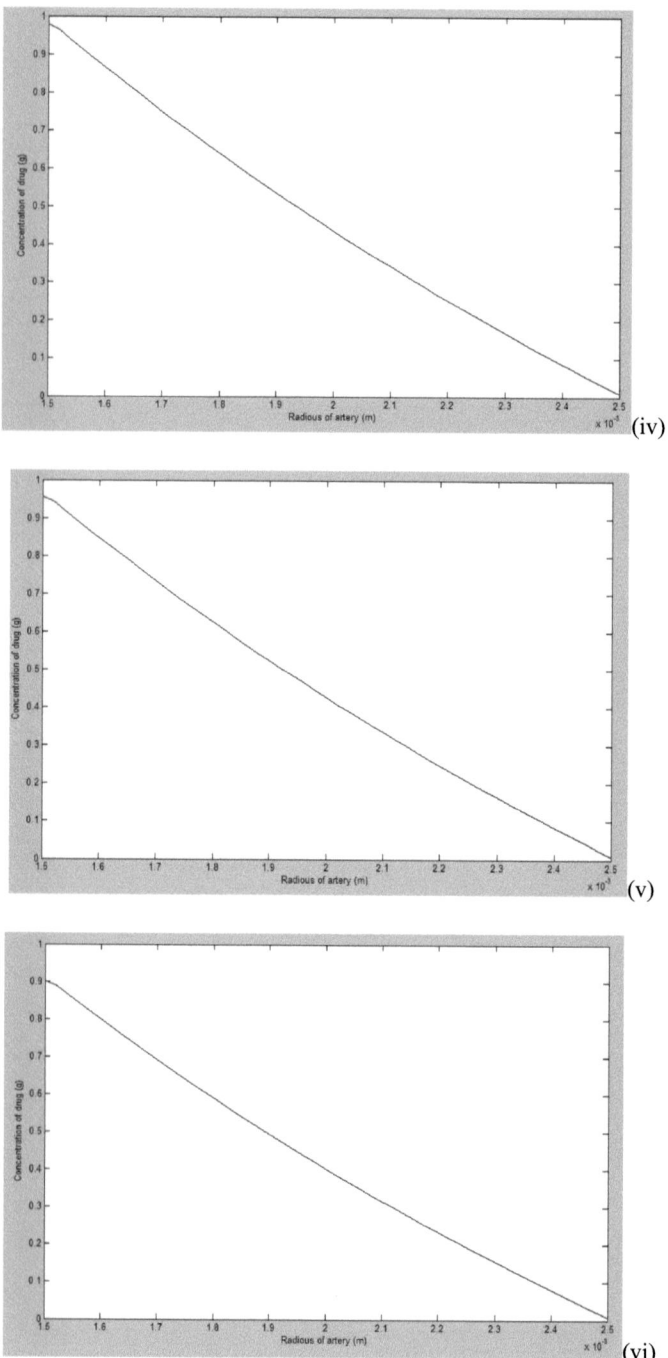

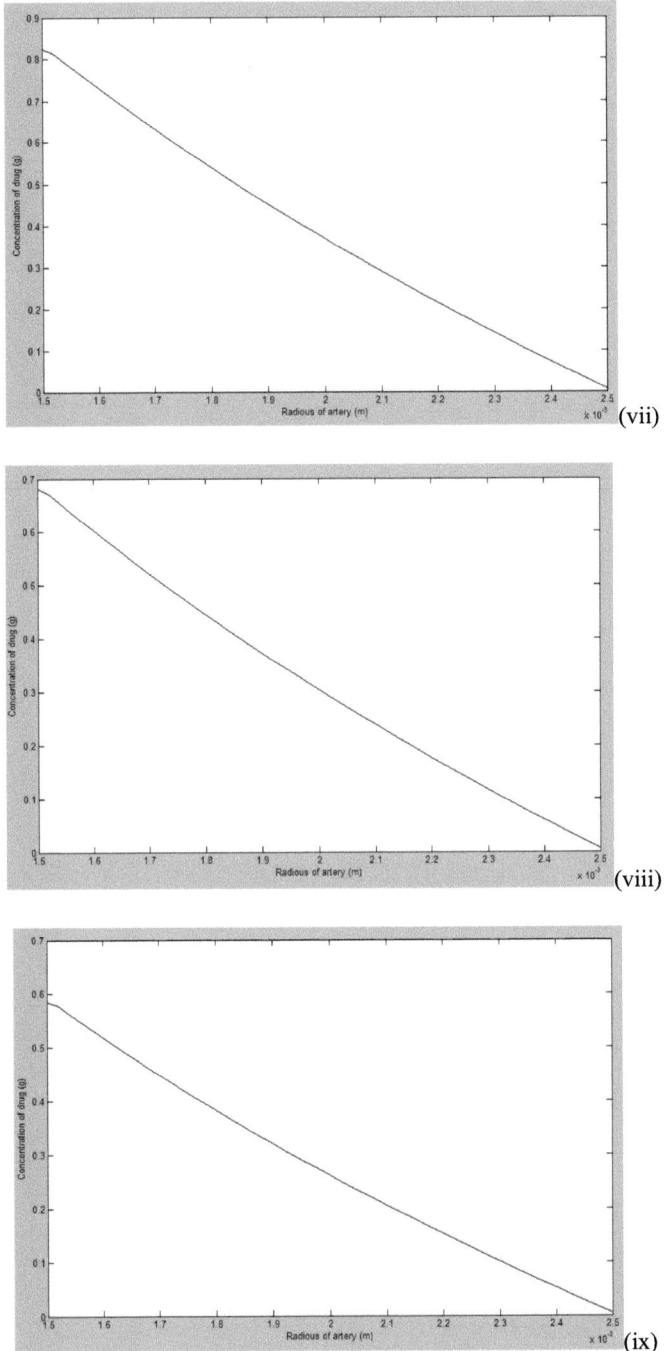

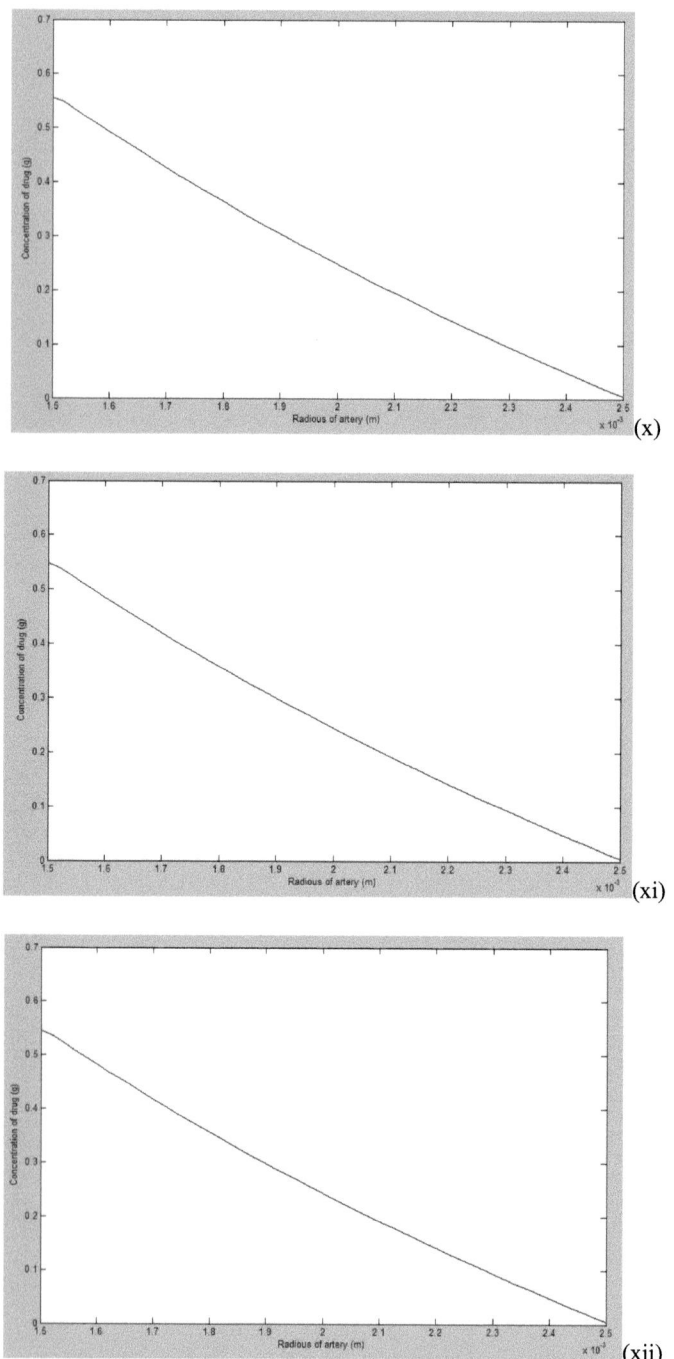

(x)

(xi)

(xii)

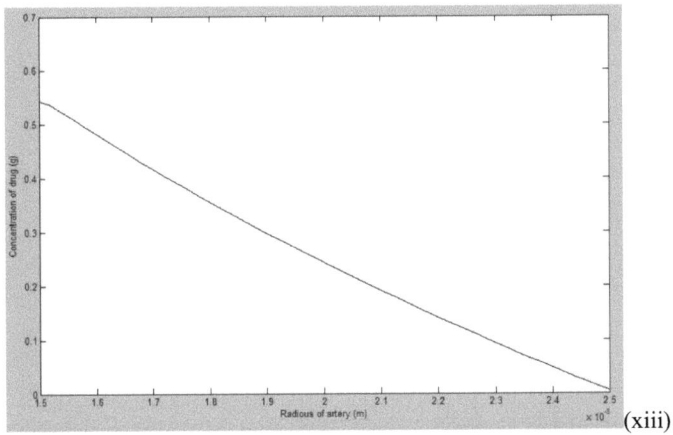
(xiii)

Fig. 5.10: Drug concentration in artery wall versus radius plot at time (i) 0 minute, (ii) 1 minute, (iii) 1 hour, (iv) 5 hours, (v) 10 hours, (vi) 1 day, (vii) 2 days, (viii) 5 days, (ix) 10 days, (x) 15 days, (xi) 20 days, (xii) 25 days and (xiii) 30 days

From these concentration versus radius plots it has been observed that drug concentration in the artery has been reduce in a non-linear fashion from inner to outer wall of the artery. T time t = 0 the drug concentration was zero at the whole wall. With increasing time at first t = 1 minute, inner artery wall drug concentration was 1.0 g/m^2 and outer wall drug concentration was 0.0. At last time, t = 30 days the inner was drug concentration was reduced to 0.5439 g/m^2 where, outer wall drug concentration was increased to 4.26115 × 10^{-3} g/m^2.

5.2.2 Drug Concentration with respect to Time

The curves representing drug concentration with respect to time at different radial are as follows.

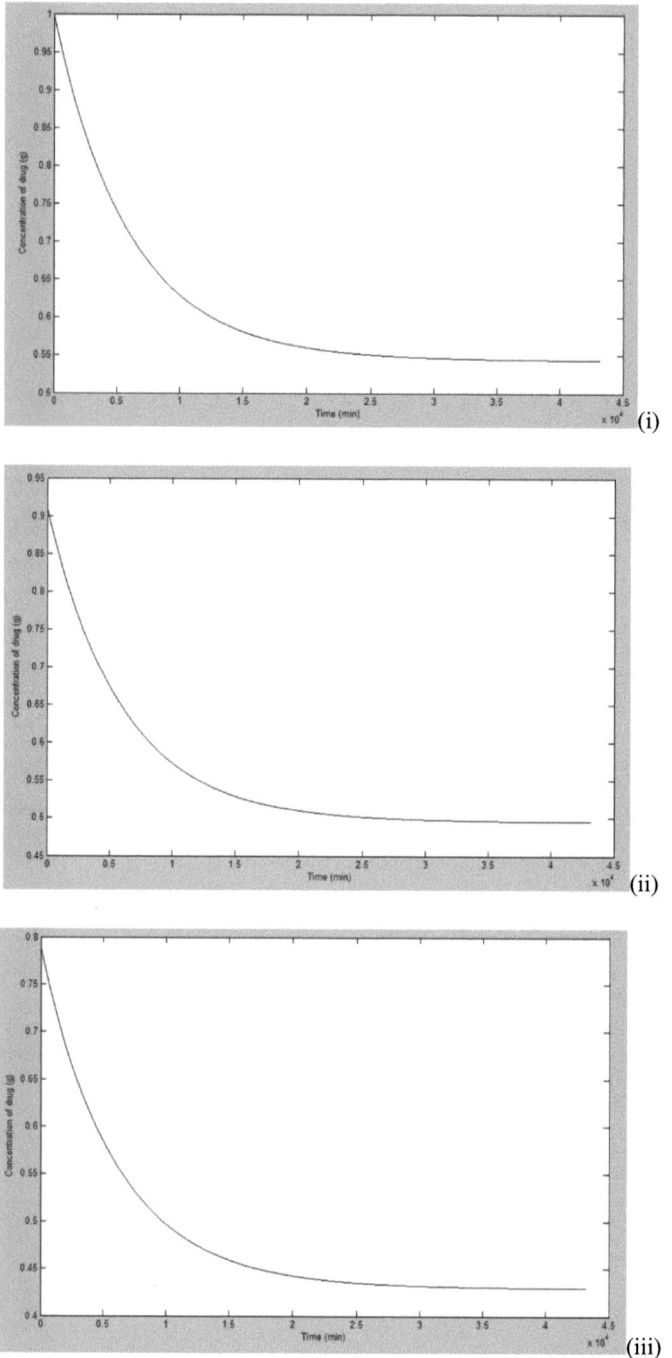

(i)

(ii)

(iii)

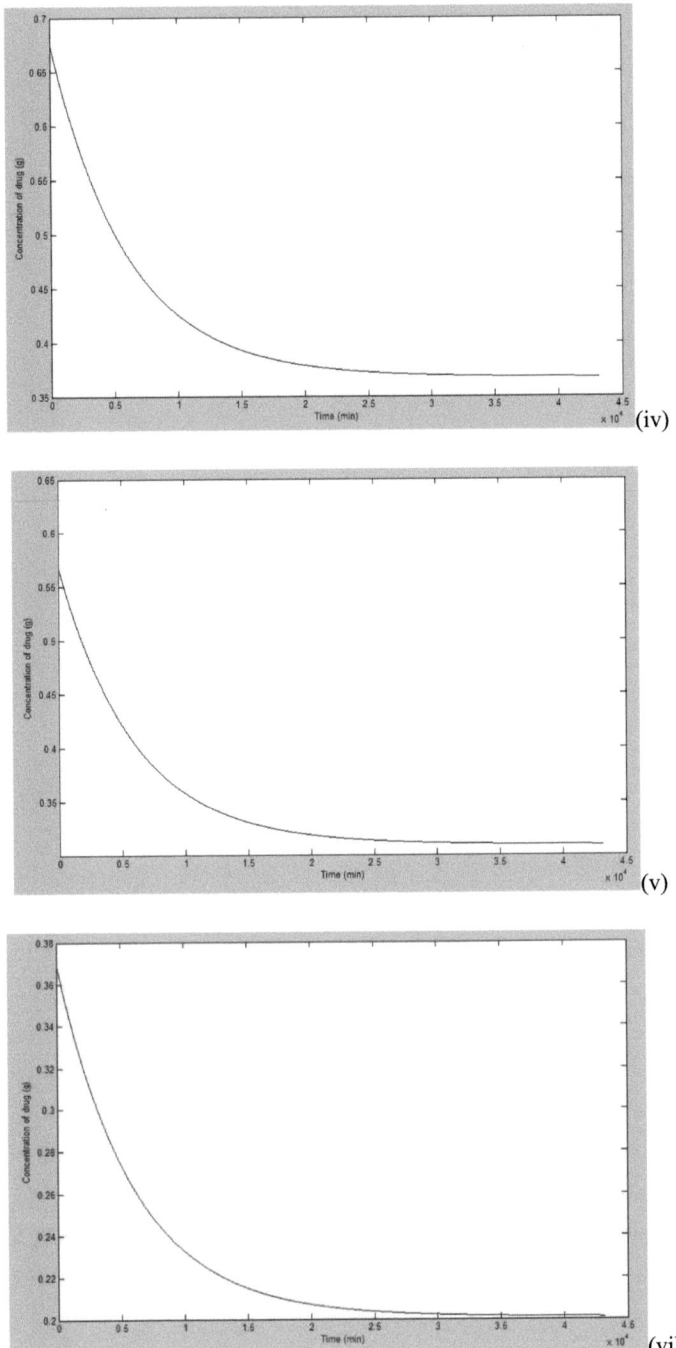

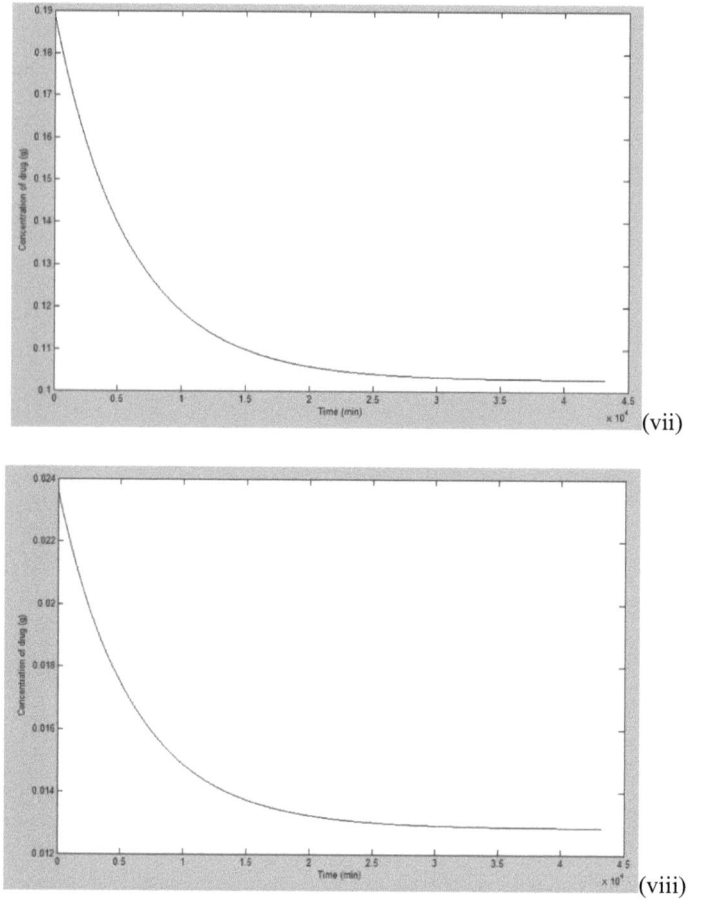

Fig.5.11: Concentration of drug at artery wall versus time at radial position (i) 1.5×10^{-3} m, (ii) 1.6×10^{-3} m, (iii) 1.7×10^{-3} m, (iv) 1.8×10^{-3} m, (v) 1.9×10^{-3} m, (vi) 2.1×10^{-3} m, (vii) 2.3×10^{-3} m and (viii) 2.5×10^{-3} m

From the above observations it has been found that, the drug concentration in the artery wall at different radial position was changed in parabolic manner. At inner artery wall drug concentration was reduced from 1.0 g/m^2 to 0.5439 g/m^2, where at the outer wall of the artery drug concentration was reduced from 0.02365 g/m^2 to 0.01285 g/m^2.

5.3 Solution of Unsteady 2D Concentration Governing Equation

The unsteady concentration equation in radial and longitudinal direction to get the drug concentration released from the DES coating at the artery wall was equation 4.7. This equation was solved using finite volume algorithm where three algebraic equation of concentration: 4.38, 4.40, 4.41 was found. Using these three equations, a tri-diagonal matrix was formed which was then solved using following MATLAB coding. In this solution procedure, DES coating thickness was taken as 12.6×10^{-6} m, Drug loading in the coating: 100 µg/cm^2 or 1 g/m^2, radius of the artery wall was varied between 1.5×10^{-3} m to 2.5×10^{-3} m, DES length was taken to be 8×10^{-3} m. Up to memory limit of MATLAB, the equation was solved for a time of 12.5 days or 18000 minutes, where discretized radius was 0.02×10^{-3} m, discretized DES length was 0.2×10^{-3} m and discretized time was 1 minute. Here diffusivity of the drug through artery wall was taken to be 1×10^{-12} m^2/s and trans-mural velocity drug through artery wall was taken to be 1×10^{-8} m/s. From this solution, the drug concentration was found in a three dimensional matrix C (t, r, z) whose dimension was $18001 \times 51 \times 40$. This matrix cannot be plotted directly of values cannot be shown in table format. MATLAB coding for this solution procedure is given in Appendix B.3

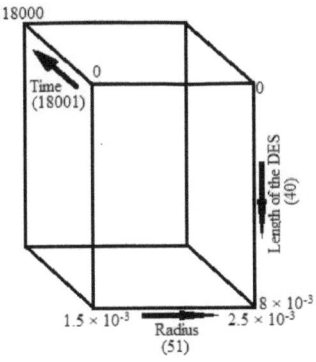

Fig. 5.12: Concentration matrix dimensions found by solving unsteady 2D governing equation

Thus by taking a certain time-radius plane (horizontal) or radius-longitudinal plane (Vertical) concentration of the drug in artery wall can be shown separately.

5.3.1 Drug Concentration Variation at Time-Radius Plane

At time-radius plane concentration profile at a certain longitudinal position is as follows:

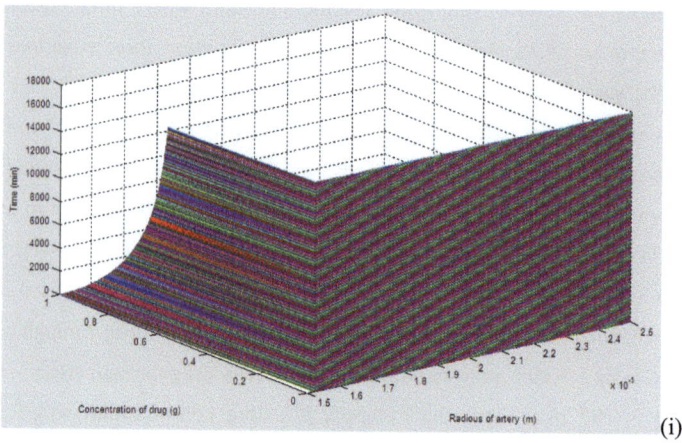
(i)

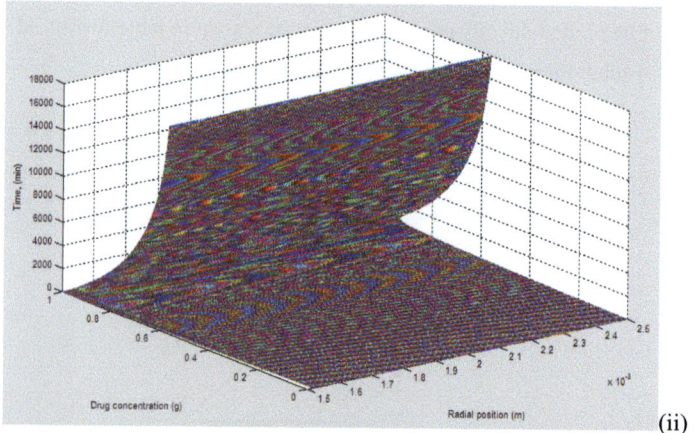
(ii)

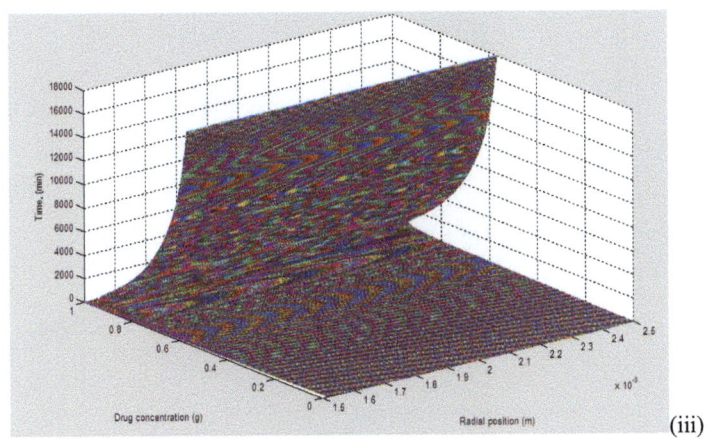
(iii)

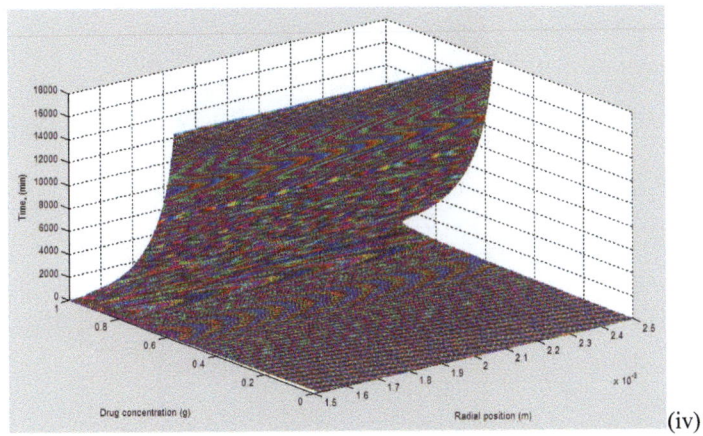
(iv)

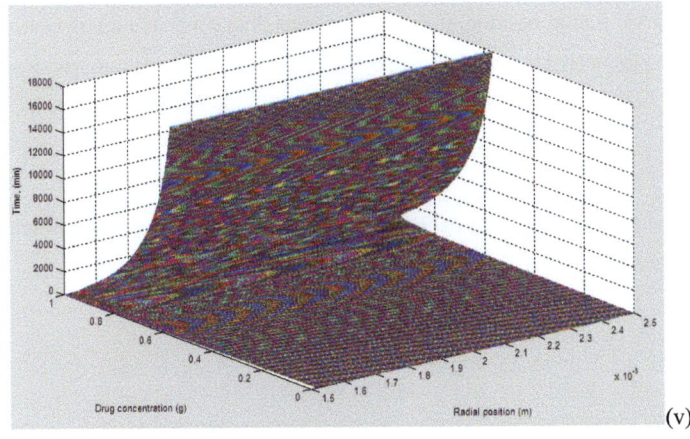
(v)

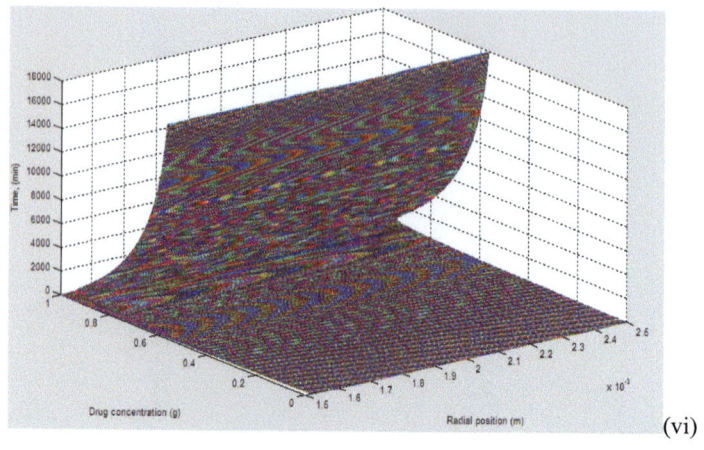

(vi)

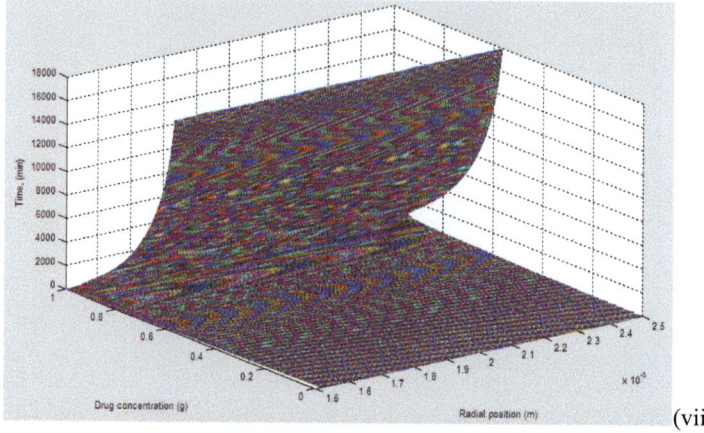

(vii)

Fig. 5.13: Radial position of the artery versus drug concentration in coating layer versus time plot at (i) $z = 0$ m, (ii) $z = 0.2 \times 10^{-3}$ m, (iii) $z = 1.0 \times 10^{-3}$ m, (iv) $z = 2.0 \times 10^{-3}$ m, (v) $z = 4.0 \times 10^{-3}$ m, (vi) $z = 6.0 \times 10^{-3}$ m and (vii) $z = 8 \times 10^{-3}$ m.

For better understanding of these plots, contour plots have been drawn, which are as follows.

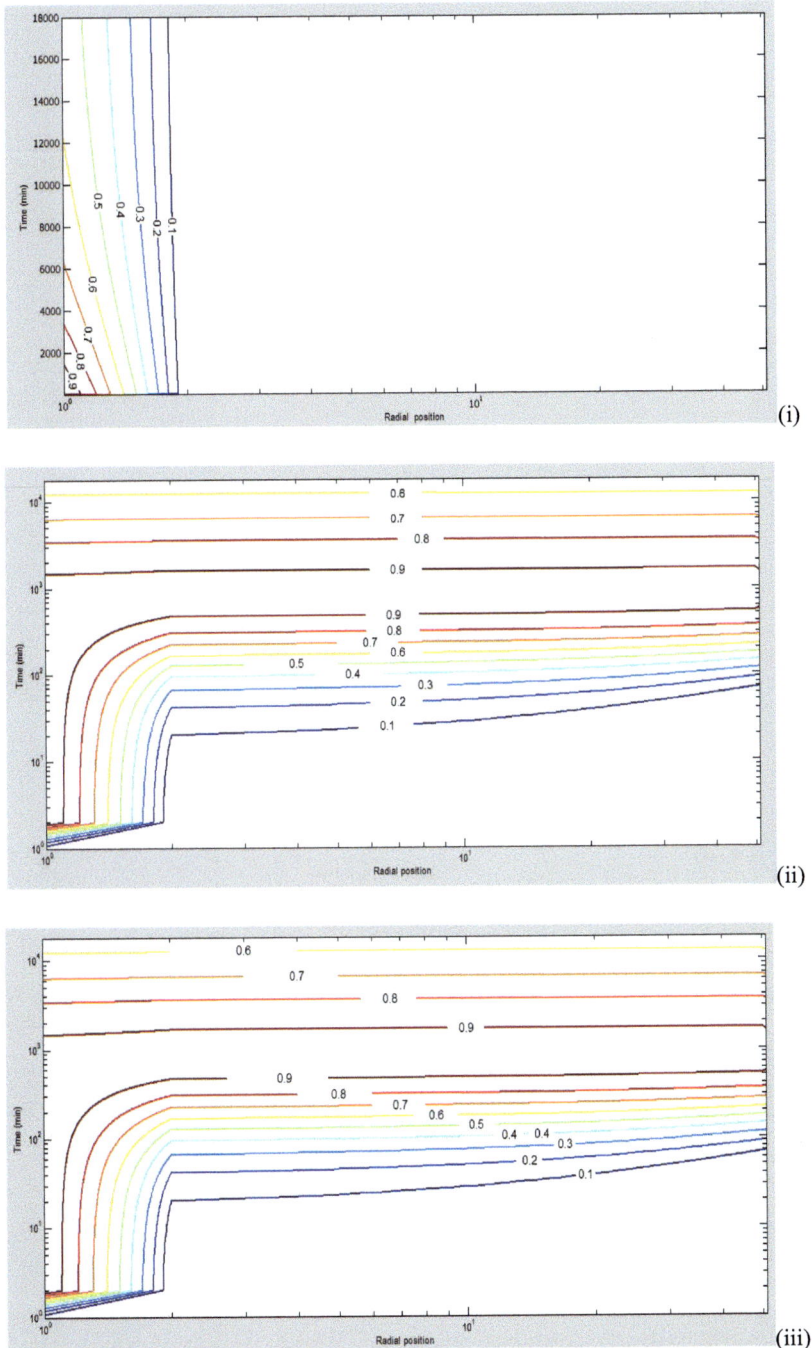

(i)

(ii)

(iii)

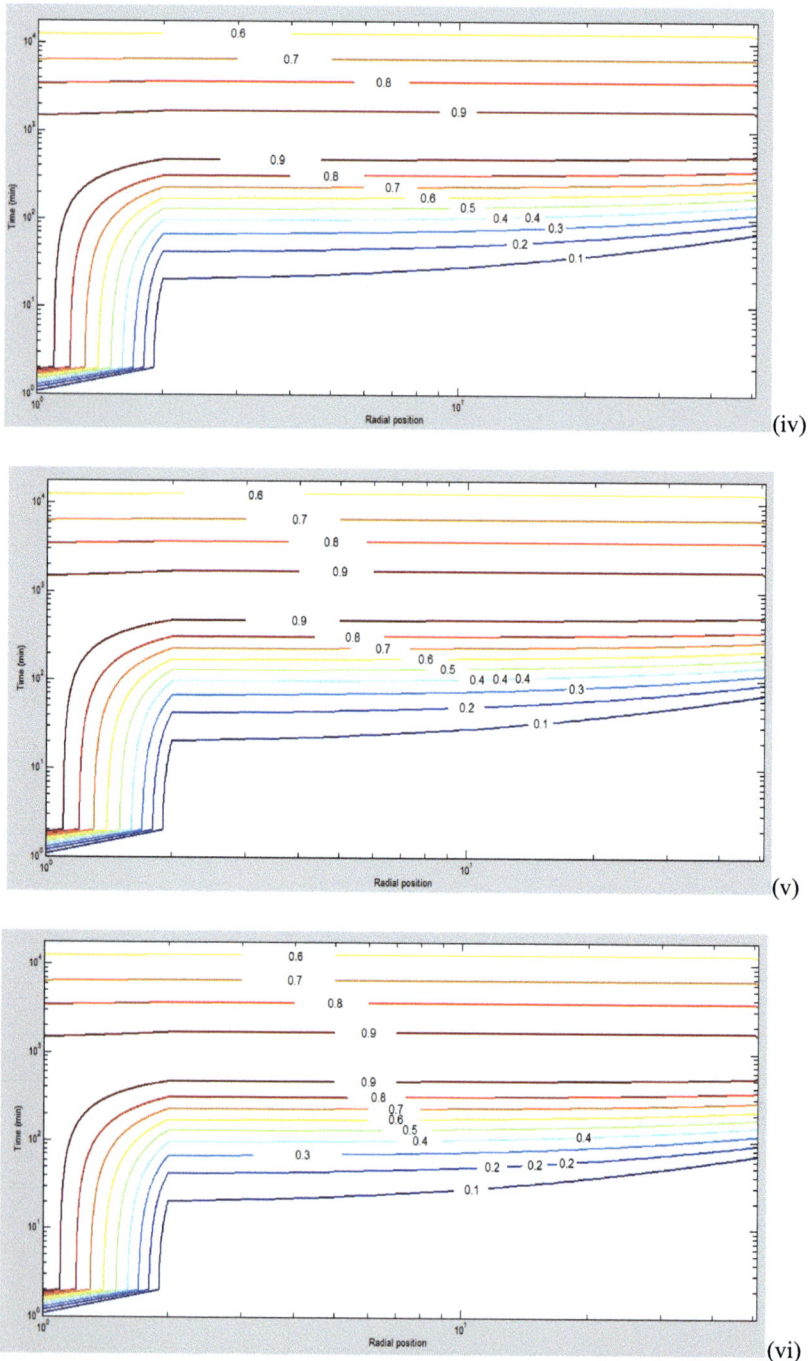

(iv)

(v)

(vi)

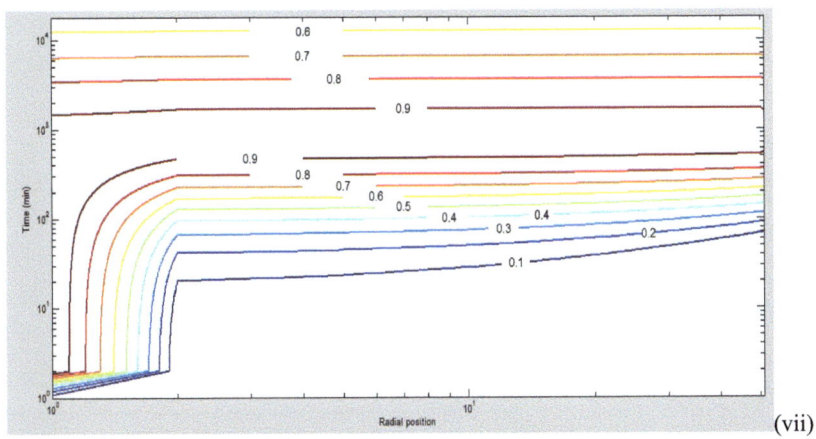

(vii)

Fig. 5.14: Contour plot of concentration of drug at artery wall at time versus radial plane at (i) z = 0 m, (ii) z = 0.2× 10^{-3} m, (iii) z = 1.0 × 10^{-3} m, (iv) z = 2.0 × 10^{-3} m, (v) z = 4.0 × 10^{-3} m, (vi) z = 6.0 × 10^{-3} m and (vii) z = 8 × 10^{-3} m. Concentration values are given in g unit.

The trend of change in drug concentration in time radius plane can be shown by drawing concentration versus radial position and concentration versus time plot. Those plots are as follows

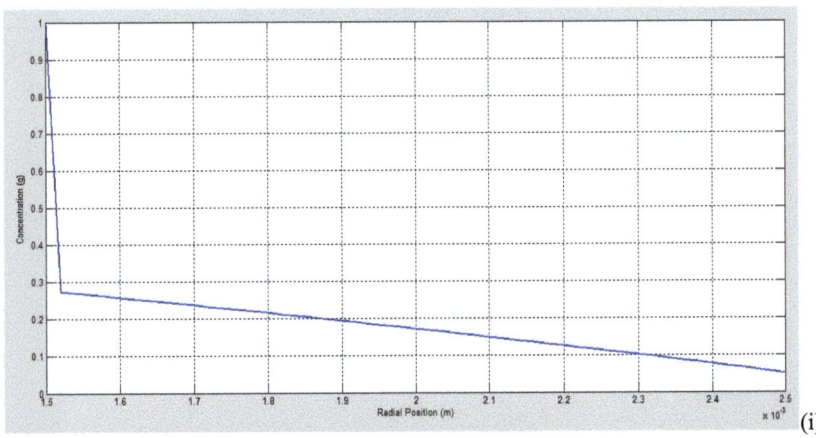

(i)

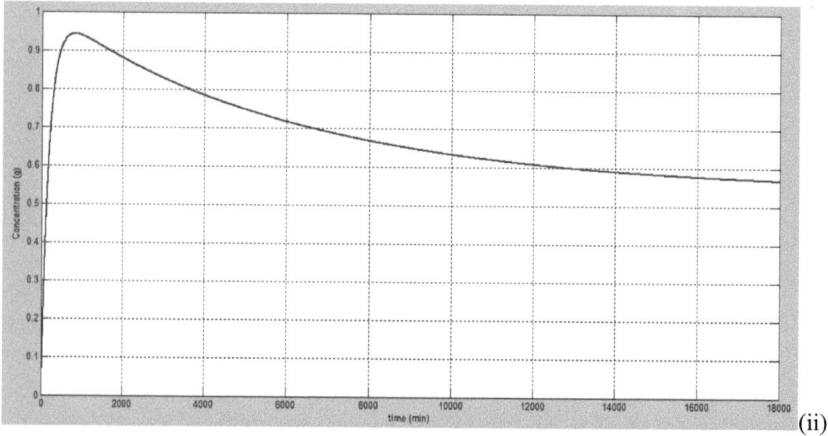

Fig. 5.15: (i) Drug concentration in artery wall versus radial position plot, (ii) drug concentration versus time plot at time radius plane at a certain z position

From the above figure 5.13 and 5.14 it is observed that at different z position, drug concentration initially increases with time and then decreases. The trend of drug concentration change with radial position is almost linear where it decreases with a steep slope at inner artery wall and with a flatter slope at the rest of the wall. The trend of drug conation change with time at a certain radial and longitudinal position is that, it increase sharply at initial period and decreases with time.

5.3.2 Drug Concentration Variation at r – z Plane

Drug concentration at r – z plane of artery wall for a certain time is as follows.

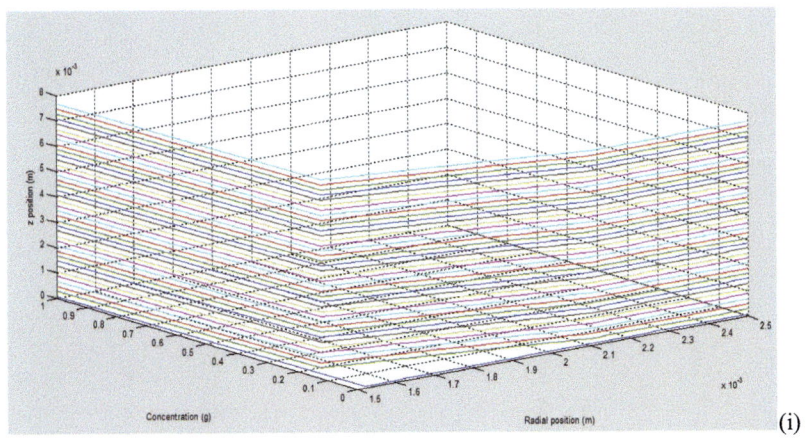

(i)

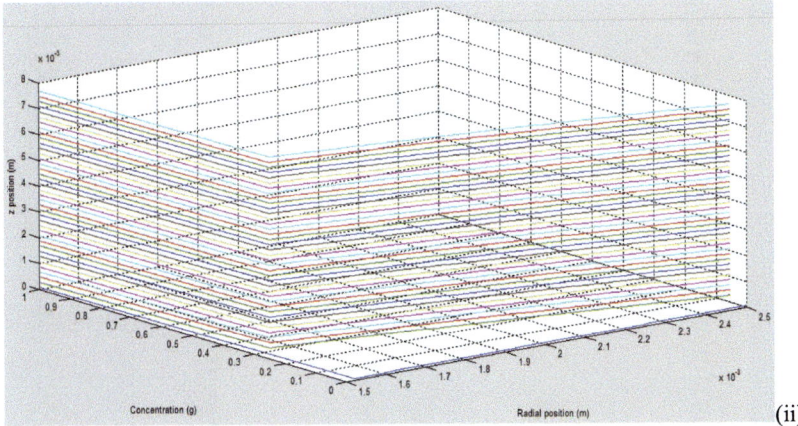

(ii)

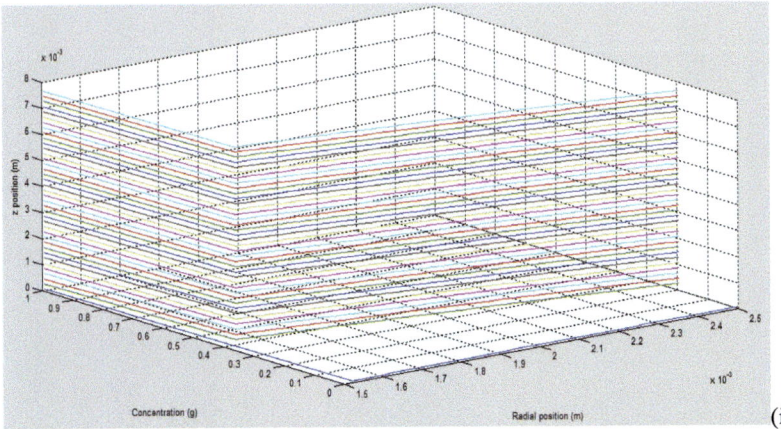

(iii)

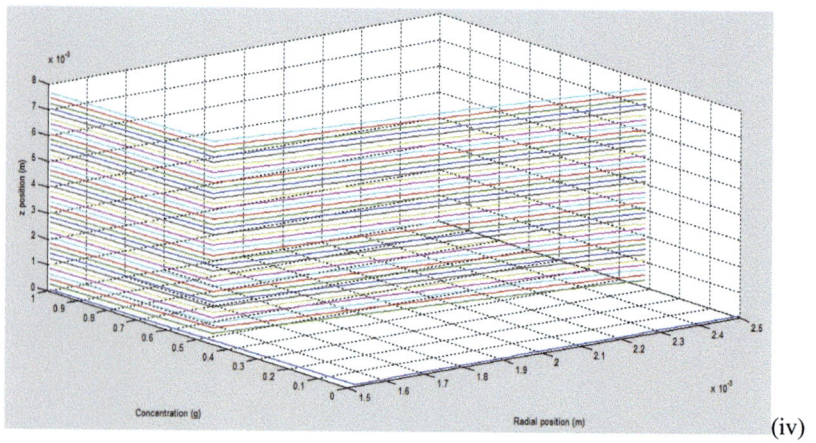

(iv)

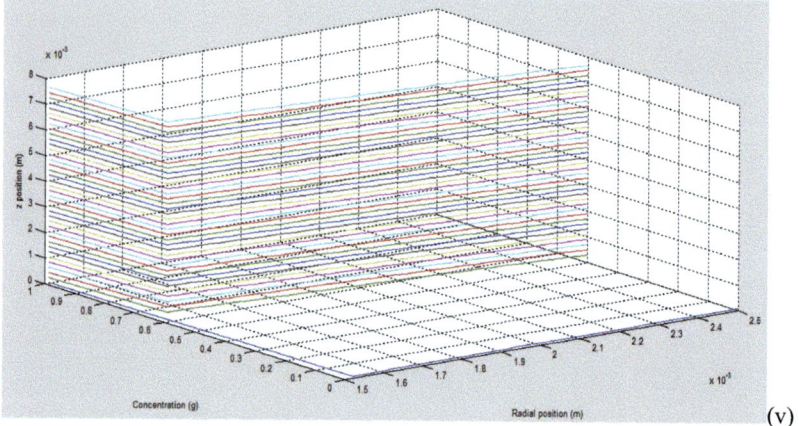

(v)

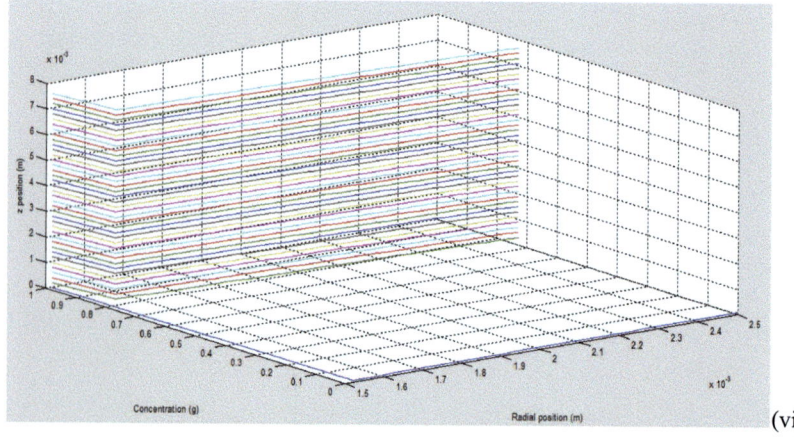

(vi)

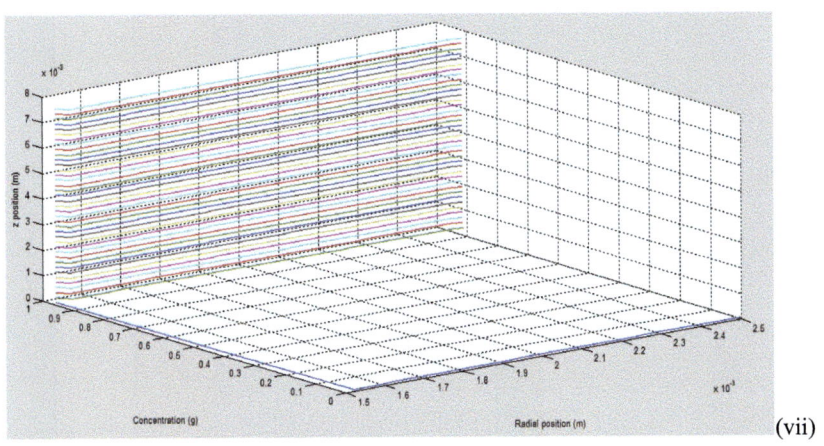
(vii)

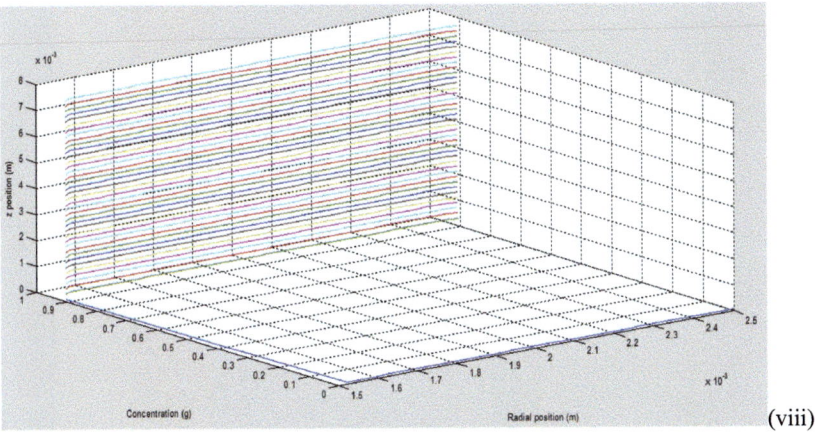
(viii)

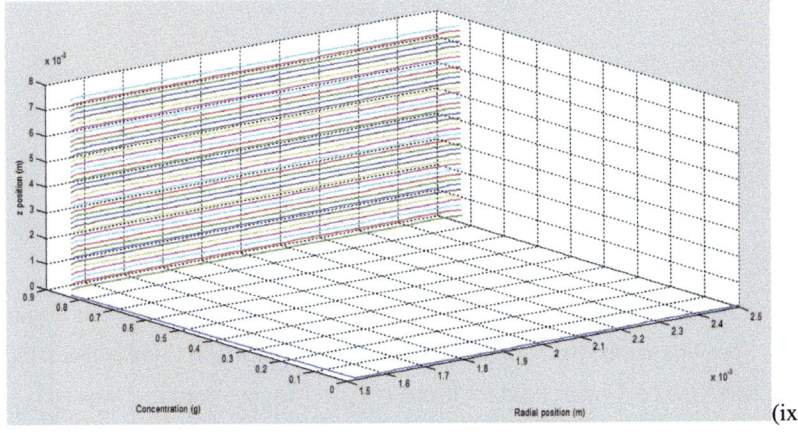
(ix)

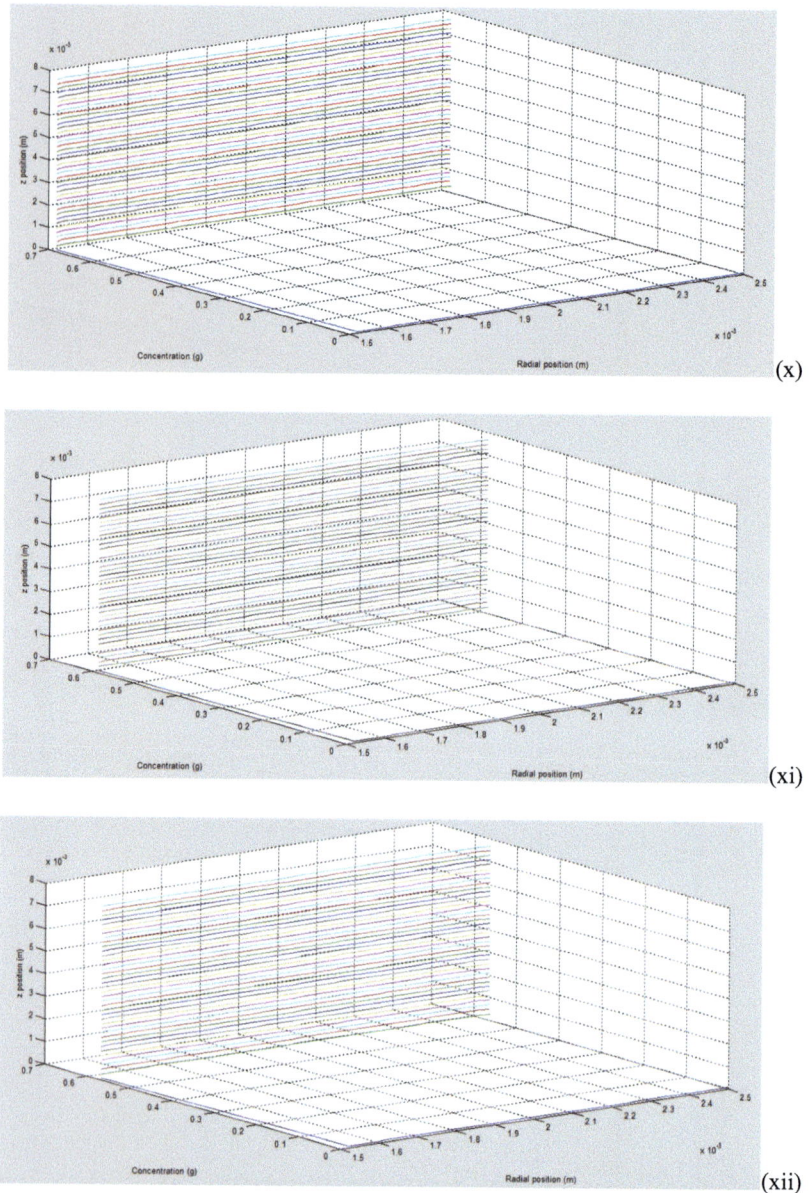

Fig. 5.16: 3D plot of Radial position versus drug concentration versus longitudinal position at time (i) t = 30 minutes, (ii) t = 60 minutes, (iii) t = 90 minutes, (iv) t = 2 hours, (v) t = 3 hours, (vi) t = 5 hours, (vii) t = 10 hours, (viii) t = 1 day, (ix) t = 2 days, (x) t = 5 days, (xi) t = 10 days and (xii) t = 12.5 days

From concentration profile given in figure 5.16 it is clearly observed that at earlier time from 30 minutes to 10 hour time range there are sharp changes of drug concentration at the radial direction at all the z position. After 10 hours, the drug concentration change along radial position become lower which gives a series of flat curves almost parallel to z axis. For proper observation of drug concentration within 10 hours time limit, contour plots of concentration in r – z plane have been drawn.

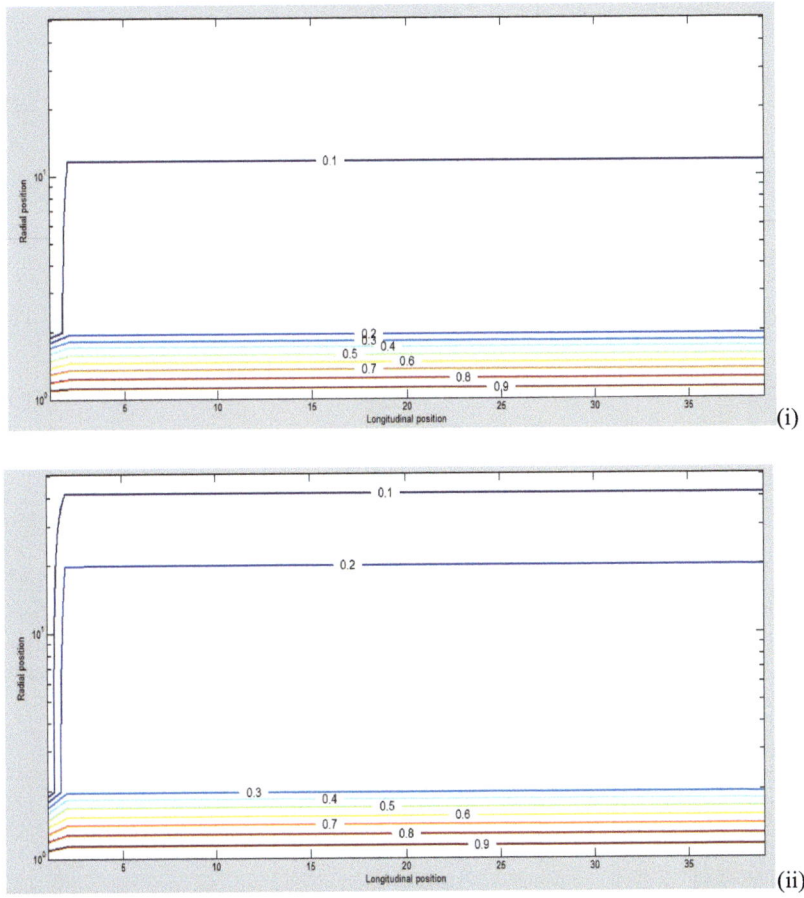

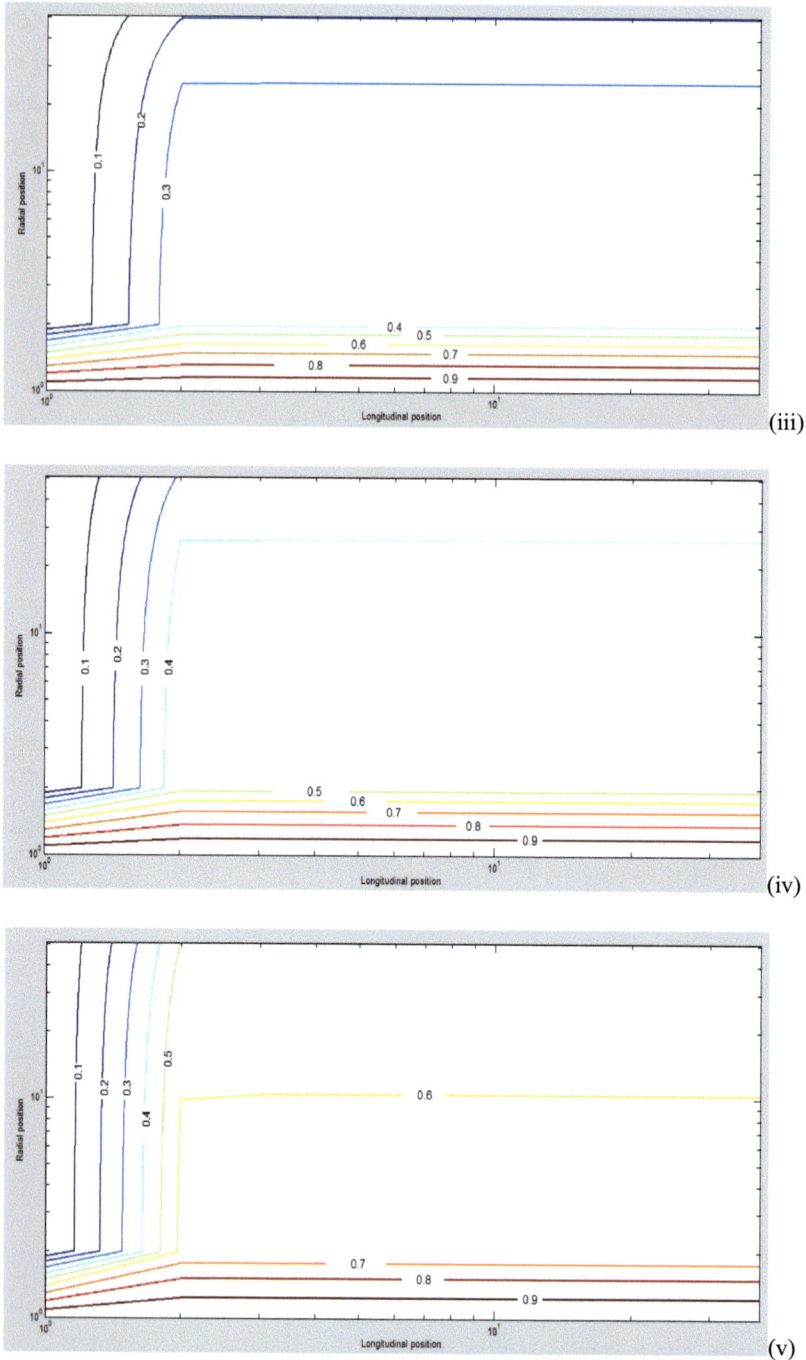

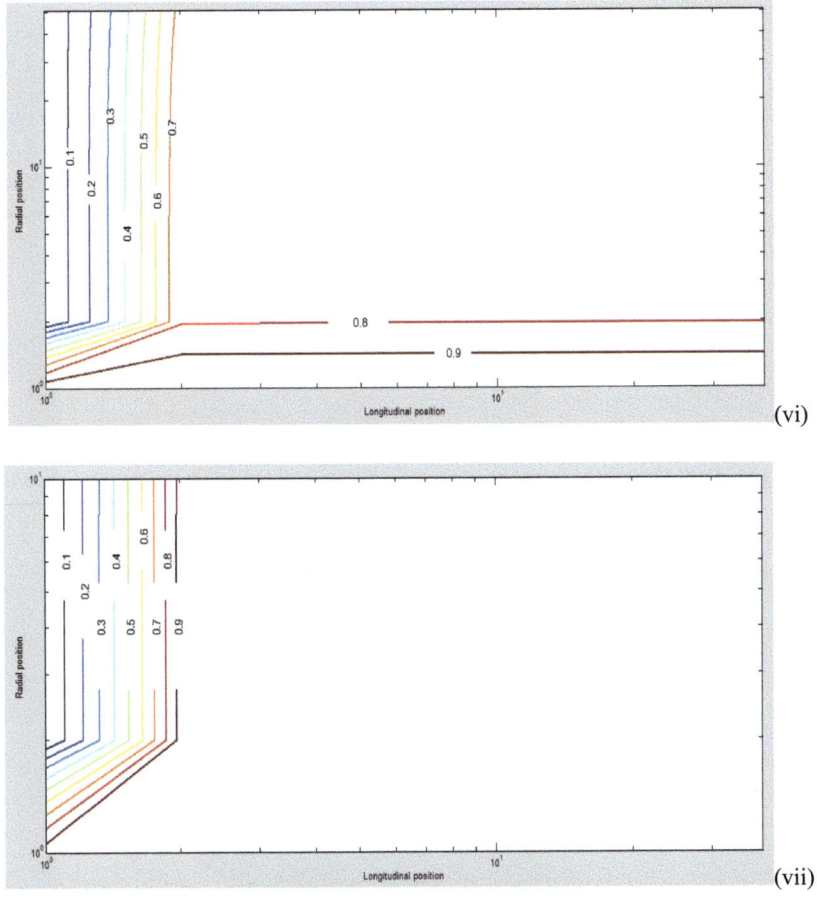

Fig. 5.17: Contour plot of concentration in z versus r plane for time (i) t = 30 minutes, (ii) t = 60 minutes, (iii) t = 90 minutes, (iv) t = 2 hours, (v) t = 3 hours, (vi) t = 5 hours, and (vii) t = 10 hours. Here concentration values are given in g unit.

To understand the trend of drug concentration change with respect to longitudinal position a concentration versus longitudinal position has been drawn at certain radial position and time stem.

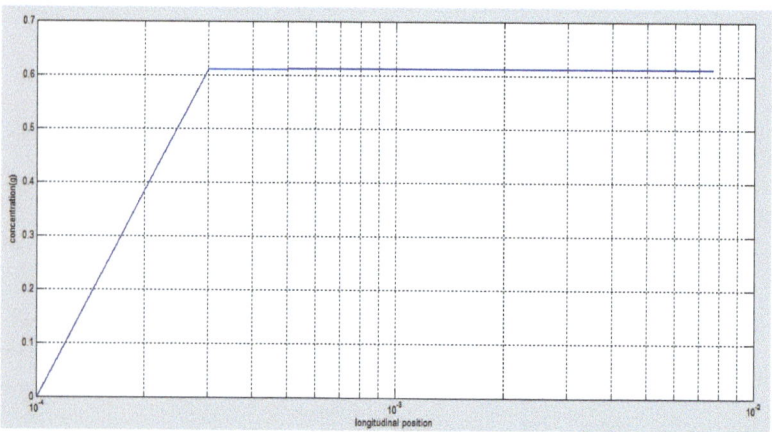

Fig. 5.18: Drug concentration at artery wall versus longitudinal position at certain radius and time step.

5.3.3 Drug Concentration Variation at time – z Plane

The variation of drug concentration at time – z plane has been shown in figurebelow.

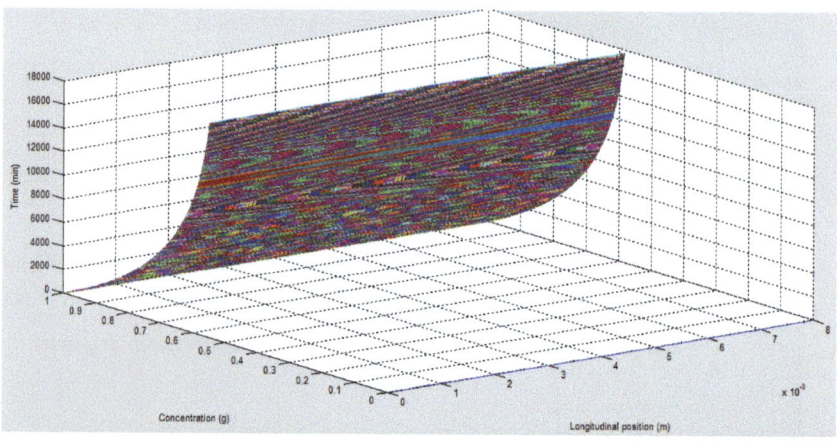

Fig. 5.19: Longitudinal position versus concentration versus time plot at inner artery wall

The change in drug concentration with respect to longitudinal position and time at inner artery wall and inside the tissue layer has been observed properly by drawing contour plot of drug concentration at time – z plane.

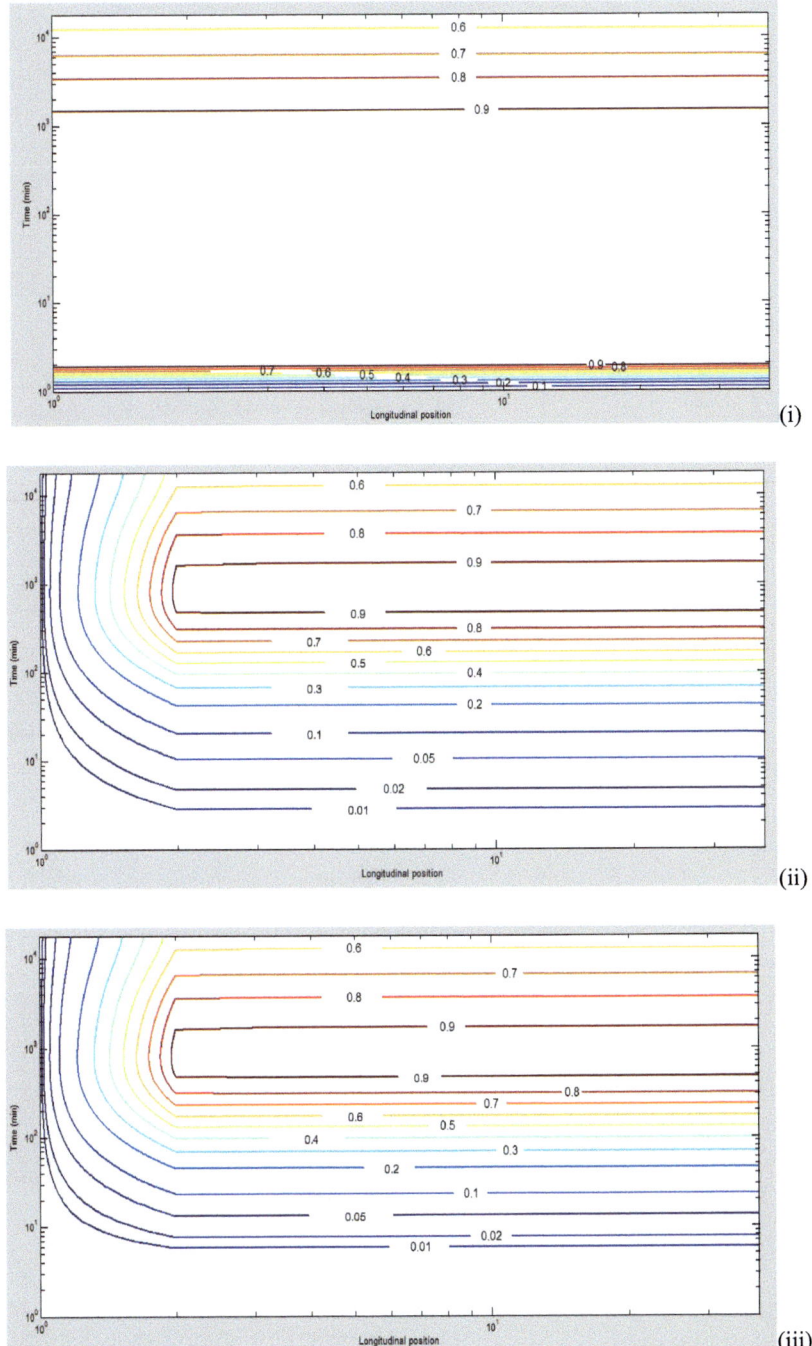

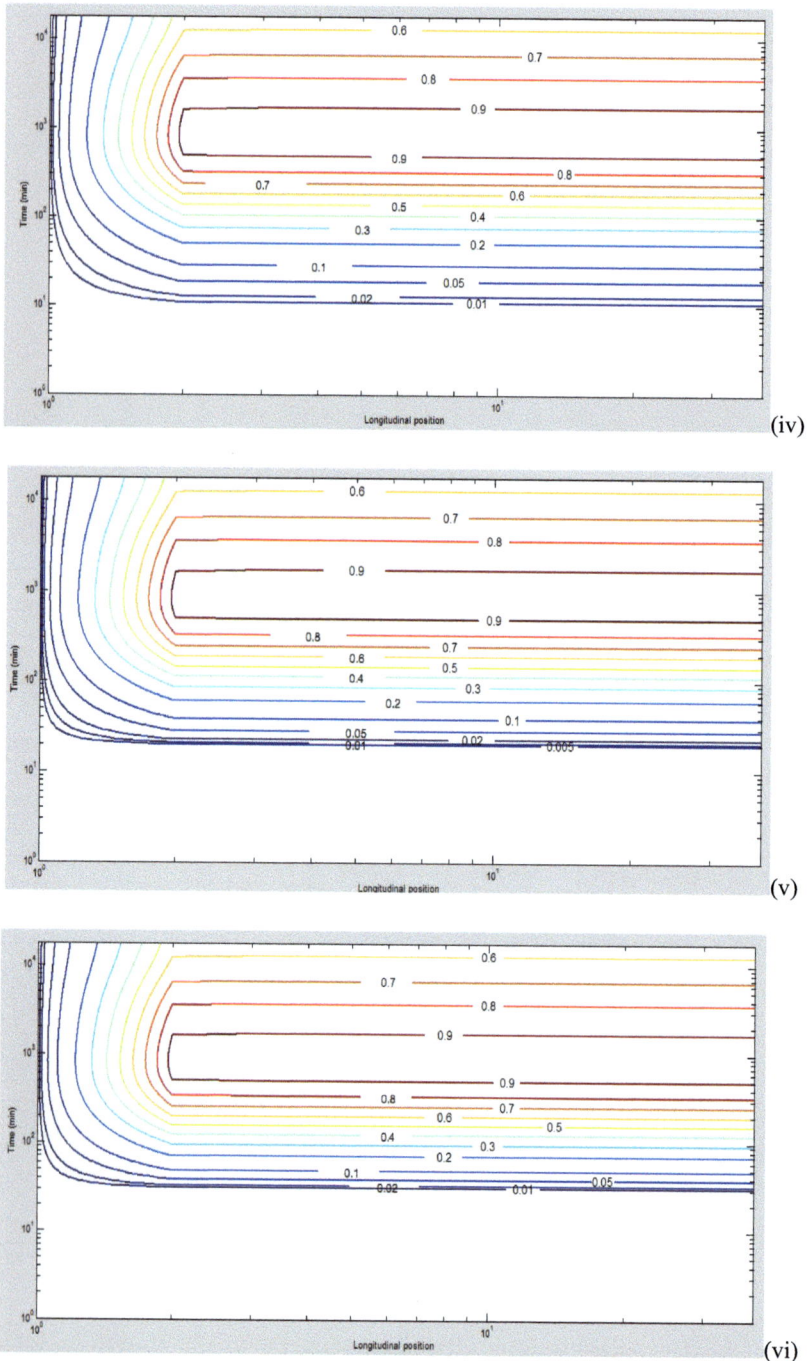

(iv)

(v)

(vi)

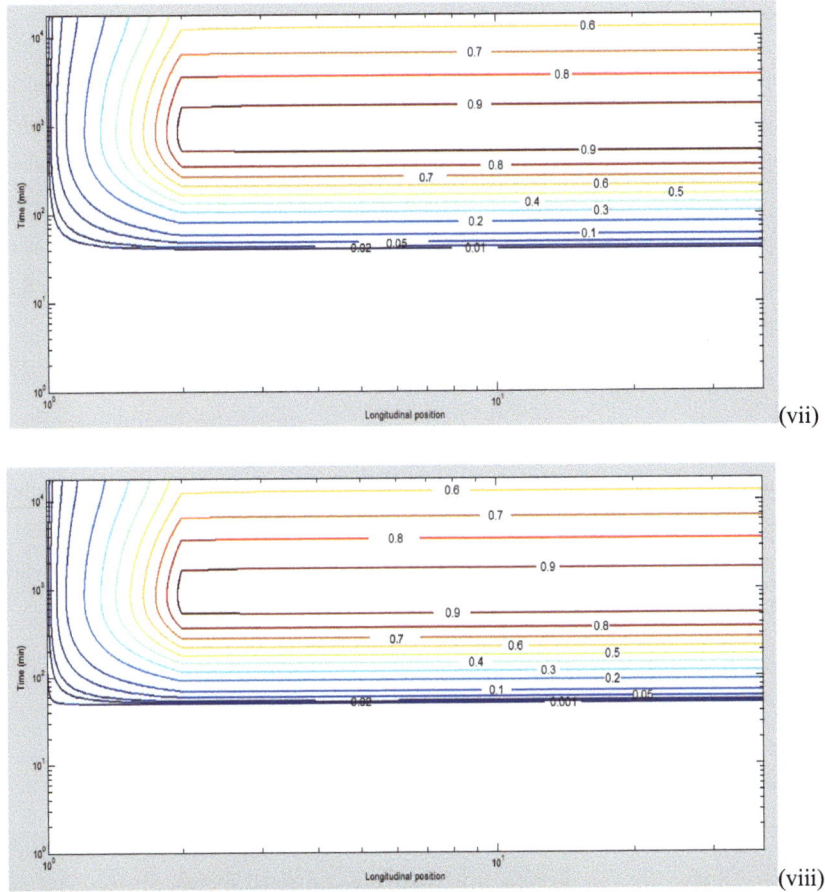

Fig. 5.20: Contour plot of drug concentration at time – z plane at (i) r = 1.50 ×10⁻³ m, (ii) r = 1.52 ×10⁻³ m, (iii) r = 1.60 ×10⁻³ m, (iv) r = 1.70 ×10⁻³ m, (v) r = 1.90 ×10⁻³ m, (vi) r = 2.10 ×10⁻³ m, (vii) r = 2.30 ×10⁻³ m and (viii) r = 2.50 ×10⁻³ m. Concentration values are taken in g unit.

From figure 5.20 it is clear that drug concentration profile shift from with respect to time when radial position is changed from 1.5×10^{-3} m to 2.5×10^{-3} m. It gives clearer insight of drug concentration profile in the artery wall.

From second model, drug concentration change in artery wall has been observed more elaborately than the first model as effect of change in longitudinal position has been incorporated here.

5.4 Grid Independency Test for the Models

In finite volume method the model is developed or any governing equation is solved depending on some grids, which are taken by considering certain discrete point interval. If a model gives almost same result for all types of grid taken than the model is considered to be the best one, is called grid independent. For example if a concentration profile is developed towards radial direction taking 0.02 grid point interval than this model will be grid independent if same concentration profile is found for 0.002 grid point interval.

5.4.1 Grid Independency Test for 1D Concentration Model

The 1D concentration model or solution of equation 4.6 has been shown in section 5.2 using grid point interval 0.02×10^{-3} m, 50 grid points were generated which gave concentration profile as shown in figure 5.8. Now taking 0.002×10^{-3} m grid interval 500 grid point is generated and then the equation is again simulated which gave following concentration profile.

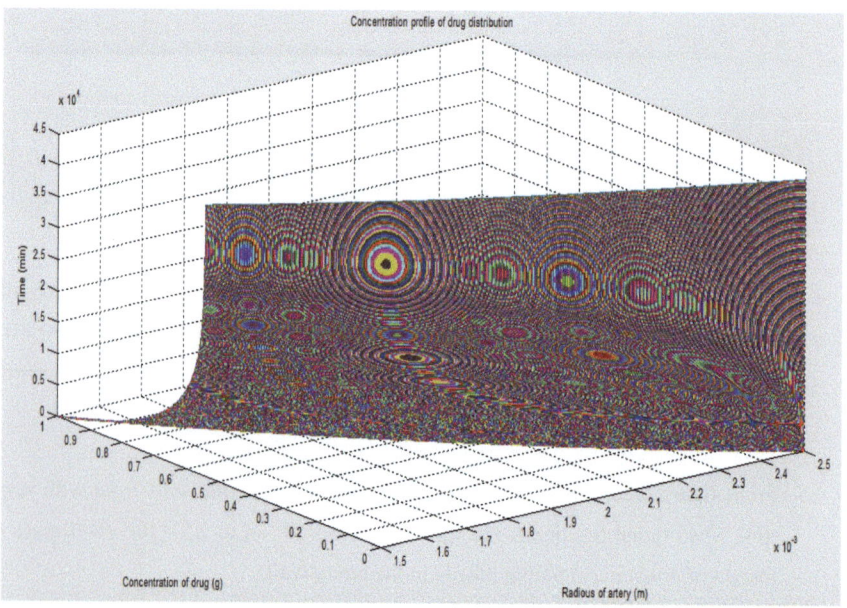

Fig. 5.21: 3D plot of radius versus drug concentration in artery wall versus time with 0.002×10^{-3} m grid interval.

If figure 5.8 and 5.21 are considered than, it is found that almost same concentration has been obtained from both grid systems. This grid independency of the model will be clearer if contour plots found from simulations are compared.

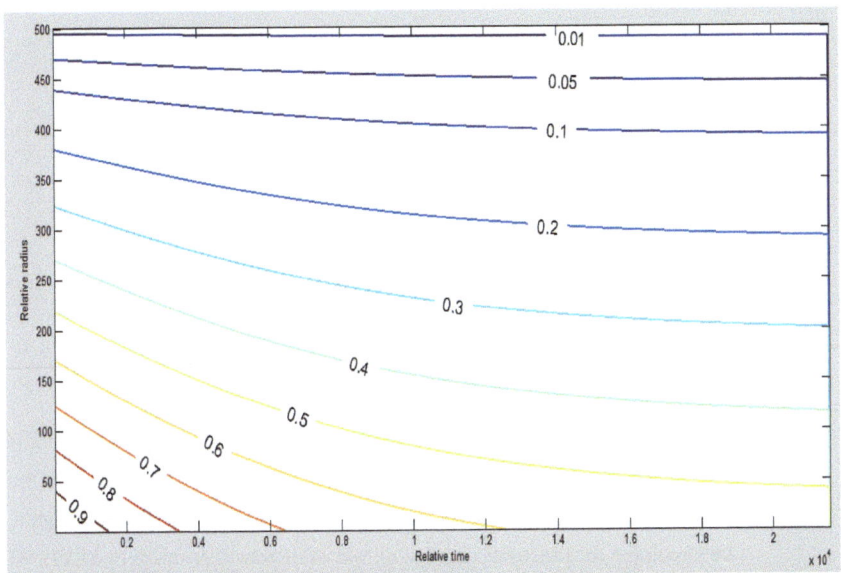

Fig. 5.22: Contour plot of concentration in relative radial and time planetime with 0.002 × 10^{-3} m grid interval. Concentration values are taken in g unit.

Comparison of figure 5.9 and 5.22 give the proof of grid independency of the model. Thus the grid system used in section 5.2 can be considered as best grid system as the second grid consumes more memory and time to simulate the model.

5.4.2 Grid Independency Test for 2D Concentration Model

The 2D concentration model or solution of equation 4.7 has been shown in section 5.3 using grid point interval 0.02 × 10^{-3} m at radial direction and 0.2 × 10^{-3} m at the longitudinal direction where 50 × 40 grids were generated, which gave concentration profile as shown in figure 5.8. Now taking 0.01 × 10^{-3} m grid interval at the radial direction and 0.1 × 10^{-3} m grid interval at longitudinal direction, 100 × 80 grids are generated and then the equation is again simulated which gave following concentration profile at z – time plane at certain radial position.

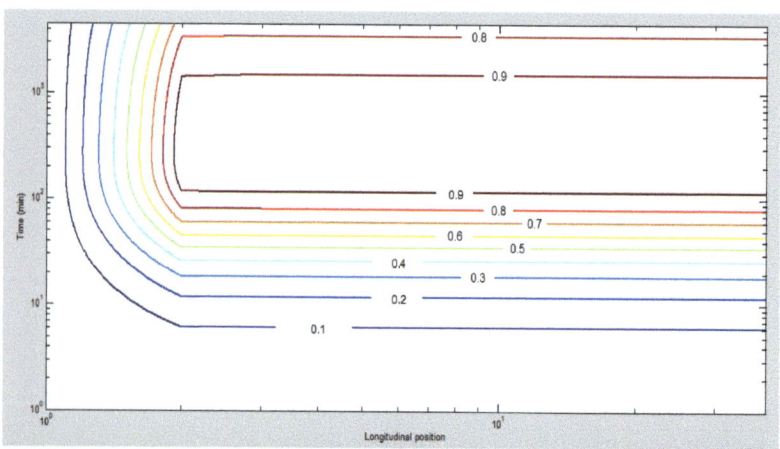

Fig. 5.23: Contour plot of drug concentration at time – z plane at r = 1.52 × 10⁻³ m

Comparison of figure 5.20 (ii) with 5.23 give the proof of grid independency of the model. Thus the grid system used in section 5.3 can be considered as best grid system as the second grid consumes more memory and time to simulate this model. Again it can be simulated for less time step due to 'out of memory' problem of MATLAB.

Chapter Six

CONCLUSION

Understanding the behavior of DES in the pursuit of improving device functionality is of great importance to clinicians and researchers alike. The modern application of computational techniques have greatly aided in achieving this goal, with researchers continuingly adding to the global understanding of drug mass transport from these devices. However, computational modeling isn't the complete solution because exact recreation ofDES deployment and ensuing mass transport is not feasible for a number of reasons. No two patients will have identical restenosis of the coronary artery and therefore a single computational model will not provide the information required to comprehensively assess the viability of a single stent design. Instead a variety of models that will cover the spectrum of DES deployment scenarios is required and to computationally recreate these like for like with*in vivo* stenting conditions would be too computationally demanding for the same rewards that one could yield with a simplified analysis. However, one must be mindful when simplifying the problem.

6.1 Conclusions

The main contribution of this thesis can be summarized as below:

- Optimization of the coating parameter n and k over DES for certain time duration (30 days) has been performed here, which will be helpful in selection of specific property coating over DES.
- Developing drug concentration profile in the wall of coronary artery with respect to time and radial position by simplifying the system as ordinary cylindrical pipe wall. Here drug concentration in the artery wall found for idealized condition.
- Developing drug concentration profile in the wall of coronary artery with respect to time, radial and longitudinal position. In this case consideration a better concentration profile was developed which reflected some realistic situation that may arise during drug delivery.

6.2 Recommendations for Future Work

There is a scope of improvement and extension of this research. The following recommendations can give some directions to the future works.

- A BSMT drug concentration governing equation was given in section 4.2 (equation 4.9-i). The possible solution procedure of that equation by finite volume algorithm has been given in section 4.3.3. This BSMT solution can be incorporated with the main drug release concentration profile developed in future.
- The model developed in this thesis work was developed by assuming drug, blood etc. to be Newtonian fluid, but in actual case all are non-Newtonian elastic fluid. By incorporating these fluid properties the present model can be modified.
- This model has been developed on the basis theoretical calculations and derivations. No experimental work was performed here. This theoretical model can be given experimental validation in future research.
- Computational models where DES struts are flush against a bare artery wall have their merits but a greater degree of complexity needs to be implemented if an improved insight into DES mass transport within the coronary artery environment is to be gained. These arteries are heavily diseased and even a thin layer of plaque between the stent strut and the wall can inhibit WSMT. The compression of the porous artery wall upon stent expansion has an interesting effect on drug concentration within the wall. The reduction in artery wall diffusivity results in higher peak concentrations beneath the stent strut can be incorporated in future concentration profile modeling.

REFERENCES

[1] Lloyd-Jones D, Adams R. Carnethon M, De Simone G, Ferguson TB, Flegal K, et al. "Heart diseases and stroke statistics – 2009 update; a report from the American Heart Association statistics committee and stroke statistics sub-committee."*Circulation*(2009); 119: 480 – 6

[2] Fischman DL, Leon MB, Baim DS, Schatz RA, Savage MP, Penn I, et al. "A randomized comparison of coronary-stent placement and balloon angioplasty in the treatment of coronary artery diseases".*N Engl. J. Med.(*1994); 331: 496 – 501

[3] Cutlip DE, Chauhan MS, Baim DS, Ho KKL, Popma JJ, Carrozza JP, et al. "Clinicalrestenosis after coronary stenting: perspectives from multicenter clinicaltrials". *J Am CollCardiol*(2002);40:2082–9

[4] Martin D., Boyle F."Drug-eluting stents for coronary artery disease: A review".*MedEngPhys* (2010), doi:10.1016/j.medengphy.2010.10.009

[5] Serruys PW, de Jaegere P, Kiemeneij F, Macaya C, Rutsch W, et al. "A comparison of balloon expandable stent implantation with balloon angioplasty in patients with coronary diseases".*N. Engl. J. Med.* (1994); 331: 489 – 95

[6] Stefanidis IK, Tolis VA, Sionis DG, Michalis LK. "Development in coronary stents".*Hel. J. Cardiol.*(2002). 43: 63 – 67

[7] Lally C, Kelly D, Prendergast P. "Stents encyclopedia of Biomedical Engineering". *Willy*. (2006)

[8] Takebayashi H, Mintz GS, Carlier SG, Kobayashi Y, Fujii K, Yasuda T, et al. "Non-uniform strut distribution correlates with more neointimal hyperplasia after sirolimus-eluting stent implantation". *Circulation* (2004); 110: 3430 – 4

[9] Ozaki Y, Violaris AG, Serruys PW. "New stent technologies".*Prog.Cardiovasc. Dis.* (1996); 39: 129 – 40

[10] Vioralis AG, Ozaki Y, Serruys PW. "Endovascular stents: a 'breakthrough technology', future challenges". *Int. J. Cardiac Imaging* (1997); 13: 3 – 13

[11] Nicholson T. "Stents: an overview". *Hosp. Med.* (1999); 60: 571 – 73

[12] Cleveland RJ, Gaines P. "Stenting inperipheral vascular diseases". *Hosp. Med.* (1999); 60: 630 – 30

[13] Barras CDJ, Myers KA. "Nitinol: its use in vascular surgery and other applications". *Eur. J. Vasc. Endovasc. Surg.* (2000); 19: 564 – 69

[14] Kastrati A, Mehilli J, Dirschinger J, Dotzer F – J, et al. "Intracoronary stenting and angiographic results: strut thickness effect on restenosis outcome (ISAR - STEREO) trial". *Circulation* (2001); 103: 2816 – 21

[15] Ross R. "The pathogenesis of atherosclerosis: a perspective for the 1990s". *Nature* (1993); 331: 489 – 95

[16] Tanaka K, Honda M, Kuramochi T. "Prominent inhibitory effects of tranilast on migration and proliferation of and collagen synthesis by vascular smooth muscle cells. *Atherosclerosis* (1994); 107: 179 – 85

[17] Serruys P, Gershlick AH. "Handbook of Drug-Eluting Stents."*Taylor& Francis* (2005)

[18] Honda Y, Grube E, de la Fuente LM, Yock PG, Stertzer SH, Fitzgerald PJ. "Novel drug-eluting stent: intravascular ultrasound observations from the first human experience with the QP2-eluting polymer stent system. *Circulation* (2001); 104: 380 – 83

[19] Liistro F, Stankovic G, Di Mario C, Takagi T, Chieffo A, et al. "First clinical experience with a paclitaxel derivative-eluting polymer stent system implantation for in-stent restenosis: immediate and long-term clinical and angiographic outcome". *Circulation* (2002); 105: 1883 – 86

[20] Degertekin M, Serruys PW, Foley DP, Tanabe K, Reger E, et al. "Persistent inhibition of neointimal hyperplasia after sirolimus-eluting stent implantation: long-term (up to 2 years) clinical, angiographic and intravascular ultrasound follow-up. *Circulation* (2002); 106: 1610 – 13

[21]　Sousa JE, Costa MA, Abizaid A, Abizaid AS, Feres F, et al. "Lack of neointimal proliferation after implantation of sirolimus-coated stents in human coronary arteries: a quantitative coronary angioplasty and three-dimensional intravascular ultrasound study" *Circle* (2001); 103: 192 – 95

[22]　Sollot SJ, Cheng L, Pauly RR, Jenkings GM, Monticone RE, et al. "Taxol inhibits neointimal smooth muscle cell accumulation after angioplasty in rat.*Clin.Interv.* (1995); 95: 1869 – 76

[23]　Bethseda, MD: Am. Soc. Health Syst. Pharm. "AHFS Drugs Information". pp. 1075 – 86. (1989)

[24]　Rowinsky EK, Donehower RC. "Drug therapy: paclitaxel (Taxol)". *N. Engl. J. Med.* (1995); 332: 1002 – 14

[25]　Yamawaki T, Shimokawa H, Kozai T, Miyata K, Higo T, et al. "Intraluminal delivery of a specific tyrosin kinase inhibitor with biodegradable stent suppresses the restenotic changes of the coronary artery in pigs in vivo". *J. Am. Coll. Cardiol.* (1998); 32: 780 – 86

[26]　Drachman DE, Edelman ER, Seifert P, Groothuis AR, Bornstain DA, et al. "Neointemal thickening after stent delivery of paclitaxel: change in composition and arrest of growth over six months". *J. Am. Coll. Cardiol.* (2000); 36: 2325 – 32

[27]　Suh H, Jeong B, Rathi R, Kim SW. "Regulation of smooth muscle cell proliferation using paclitaxel-loaded poly (ethylene oxide)-poly (lactide/ glycolide) nanospheres". *J. Biomed. Mater. Res.* (1998); 42: 331 – 38

[28]　Alexis F, Venkatraman SS, Rath SK, Boye F. "In vitro study of release mechanisms of paclitaxel and rapamycin from drug-incorporated biodegradable stent matrices". *J. Control. Release* (2004); 98: 67 – 74

[29]　Burton PBJ, Yacoub MH, Barton PJR. "Rapamycin (sirolimus) inhibits heart cell growth in vitro". *Pediatr.Cardiol.* (1998); 19: 468 – 70

[30]　Kahan BD, Camardo JS. "Rapamycin: clinical result and future opportunities". *Transplantation.* (2001); 72: 1181 – 93

[31]   Morice MC, Serruys PW, Sousa JE, Perin M, Colombo A, et al. "RAVEL study group, a randomized comparison of a sirolimus-eluting stent with a standard stent for coronary revascularization". *N. Engl. J. Med.* (2002); 346: 1773 – 80

[32]   Regar E, Laarman G, Blanchard D, Eltchaninoff H, Sousa JE, et al. "Sirolimus, inhibits restenosis irrespective of the vessel size: a subanalysis of multicenter RAVEL trial". *J. Am. Coll. Cardiol.* (2002); 39: 58A

[33]   Degertekin M, Regar E, Tanabe K, Smits PC, van-der-Gissen WZ, et al. "Sirolimus-eluting stent for the treatment of complex in-stent restenosis, the first clinical experiences". *J. Am. Coll. Cardiol.* (2003); 41: 184 – 89

[34]   Rager E, Serruys PW, Bode C, Holubarsch C, Guermonprez JL, et al. "Angiographic findings of the multicenter randomized study with the sirolimus-eluting Bx-velocity balloon expandable stent (RAVEL): sirolimus-eluting stents inhibit restenosis irrespective of the vessel size". *Circulation* (2002); 106: 1949 – 56

[35]   Moses JW, Leon MB, Popma JJ, Fitzgerald PJ, Holmes DR, et al. "Sirolimus-eluting stents versus standard stents in patients with stenosis in a native coronary artery. *N. Engl. J. Med.* (2003); 349: 1315 – 23

[36]   Grube E, Silber S, Hauptmann KE, Muller R, Buellesfeld L, et al. "TAXUS I: six and twelve month results from a randomized, double-blind trial on a slow-release paclitaxel-eluting stent for de novo coronary lesions". *Circulation* (2003); 107: 38 – 42

[37]   Tanabe K, Serruys PW, Deggertekin M, Guagliumi G, Grube E, et al. "Chronic arterial responses to polymer controlled paclitaxel-eluting stents: comparison with bare metal stents by serial intravascular ultrasound analysis: data from the randomized TAXUS-II trial". *Circulation* (2004); 109: 196 – 200

[38]   Tanabe K, Serruys PW, Grube E, Smits PC, Selbach G, et al. "TAXUS III trial: in-stent restenosis treated with stent based delivery of paclitaxel incorporated in slow release polymer formulation". *Circulation* (2003); 107: 559 – 64

[39] Stone GW, Ellis SG, Cox DA, Hermiller J, O'Shaughnessy C, et al. "A polymer-based, paclitaxel-eluting stent in patients with coronary artery diseases" *N. Engl. J. Med.* (2004); 350: 221 – 31

[40] Tsuji T, Tamai H, Igaki K, Kyo E, Kosuga K, et al. "Biodegradable stents as a platform to drug loading". *Int. J. Cardiovasc. Interv.* (2003); 5: 13 -16

[41] Moses JW, Leon MB, PopmaJJ, Fitzgerald PJ, Holmes DR, O'Shaughnessy C, et al. "Sirolimus-eluting stents versus standard stents in patients with stenosis in anative coronary artery". *N. Engl. J. Med.*(2003);349: 1315 – 23

[42] Sousa JE, Costa MA, Abizaid A, Abizaid AS, Feres F, Pinto IMF, et al. "Lack of neointimal proliferation after implantation of sirolimus-coated stents inhuman coronary arteries: a quantitative coronary angiography and threedimensionalintravascular ultrasound study. *Circulation*(2001);103:192–5

[43] Morice M-C, Serruys PW, Sousa JE, Fajadet J, Ban Hayashi E, Perin M, et al. "Arandomized comparison of a sirolimus-eluting stent with a standard stent forcoronary revascularization". *N. Engl. J. Med.*(2002);346: 1773 – 80

[44] Schofer J, Schlüter M, Gershlick AH, Wijns W, Garcia E, Schampaert E, et al. "Sirolimus-eluting stents for treatment of patients with long atheroscleroticlesions in small coronary arteries: double-blind, randomized controlled trial(E-SIRIUS)". *Lancet*(2003);362: 1093 – 9

[45] Schampaert E, Cohen EA, Schluter M, Reeves F, Traboulsi M, Title LM, et al. "The Canadian study of the sirolimus-eluting stent in the treatment of patients with long de novo lesions in small native coronary arteries (C-SIRIUS)". *J. Am. Coll.Cardiol.*(2004);43: 1110 – 5

[46] Meital Z, Robert CE. "Drug-eluting bioresorbable stents for various application". *Annu.Rev. Biomed. Eng.* (2006); 8: 153 – 80

[47] Chen MC, Liang HF, Chinu YL, Chang Y, Wei HJ, Sung HW. "A novel drug-eluting stent spray coated with multilayer collagen and sirolimus". *J. Control. Release* (2005); 108: 178 – 189

[48] Colombo A, Drzewiecki J, Banning A, Grube E, Hauptmann K, Silber S, et al. "Randomized study to assess the effectiveness of slow- and moderate-release polymer-based paclitaxel-eluting stents for coronary artery lesions". *Circulation*(2003);108: 788 – 94

[49] Turco MA, Ormiston JA, Popma JJ, Mandinov L, O'Shaughnessy CD, Mann T,et al., McGarry TF. "Polymer-based, paclitaxel-eluting TAXUS Libertestentin de novo lesions: the pivotal TAXUS ATLAS trial". *J. Am. Coll.Cardiol.* (2007);49:1676–83

[50] Meredith IT, Ormiston J, Whitbourn R, Kay P, Muller D, Bonan R, et al. "First-in-human study of the Endeavor ABT-578-eluting phosphoryl-cholineen-capsulatedstent system in de novo native coronary artery lesions:endeavor I trial". *Euro Intervention*(2005);1: 157 – 64

[51] Fajadet J, Wijns W, Laarman G-J, Kuck K-H, Ormiston J, Munzel T, et al.Randomized, double-blind, multicenter study of the endeavor zotarolimus-elutingphosphoryl-chlorine-encapsulated stent for treatment of native coronary artery lesions: clinical and angiographic results of the ENDEAVORII trial. *Circulation*(2006);114:798–806

[52] Kandzari DE, Leon MB,PopmaJJ, Fitzgerald PJ, O'Shaughnessy C, BallMW,et al. "Comparison of zotarolimus-eluting and sirolimus-eluting stents in patients with native coronary artery disease: A randomized controlled trial". *J. Am. Coll. Cardiol.*(2006);48:2440–7

[53] Leon MB, Mauri L, Popma JJ, Cutlip DE, Nikolsky E, O'Shaughnessy C, et al. "Arandomized comparison of the Endeavor zotarolimus-eluting stent versus the TAXUS paclitaxel-eluting stent in de novo native coronary lesions: 12-monthoutcomes from the ENDEAVOR IV trial". *J. Am. Coll.Cardiol.*(2010);55:543– 54

[54] Kandzari DE, Mauri L, Popma JJ, Turco M, O' Shaunessy C, Gurbel PA, et al. "ENDEAVOR III: 5 year final outcomes".Presented at The AmericanCollege of Cardiology Scientific Sessions, 2010.

[55] Leon MB. "TheENDEAVORandENDEAVOR resolute zotarolimus-elutingstent:comprehensive update of the clinical trial program". Presented at:Trans-catheter Cardiovascular Therapeutics, 2009.

[56] Serruys PW, Ong ATL, Piek JJ, Neumann F-J, van der Giessen WJ, Wiemer M, et al. "A randomized comparison of a durable polymer Everolimus-elutingstent with a bare metal coronary stent: the SPIRIT first trial". *Euro Intervention* (2005);1:58–65

[57] Serruys PW, Ruygrok P, Neuzner J, Piek JJ, Seth A, Schofer J, et al. "Arandomizedcomparison of an everolimus-eluting coronary stent with a paclitaxel-eluting coronary stent: the SPIRIT II trial. *Euro Intervention*(2006);2:286–94

[58] Stone GW, Midei M, Newman W, Sanz M, Hermiller JB, Williams J, et al. "Comparison of an Everolimus-eluting stent and a Paclitaxel-eluting stent in patients with coronary artery disease: a randomized trial". *J. Am. Med. Assoc.* (2008);299:1903–13

[59] Stone GW, Rizvi A, Newman W, Mastali K, Wang JC, Caputo R, et al. "Everolimus-eluting versus paclitaxel-eluting stents in coronary artery disease.*N. Engl. J. Med.*(2010);362:1663–74

[60] Garg S, Serruys P, Onuma Y, Dorange C, Veldhof S, Miquel-Hebert K, et al. "3-Year clinical follow-up of the XIENCE V Everolimus-eluting coronary stent system in the treatment of patients with de novo coronary artery lesions: The SPIRIT II trial (Clinical evaluation of the Xience V Everolimus eluting coronary stent system in the treatment of patients with de novo native coronary artery lesions)". *J. Am. Coll.Cardiol.*(2009);2:1190 – 8

[61] Stone GW. "The XIENCE V-PROMUS everolimus-eluting stent:comprehensiveupdate of the clinical trial program". Presented at: Trans-catheter Cardiovascular Therapeutics, 2009

[62] Abizaid A, Chan C, Lim Y-T, Kaul U, Sinha N, Patel T, et al. "Twelve-month outcomes with a paclitaxel-eluting stent transitioning from controlled trials to clinical practice (the WISDOM registry)". *Am. J. Cardiol.* (2006); 98: 1028 – 32

[63] Kimura T, Yokio H, Nakagawa Y. "Three year follow-upafter implantation of metallic coronary artery stents". *N, Rngl. J. Med.* (1996); 334: 561 – 66

[64] Eberhart RC, Su SH, Kytai TN, Zilberman M, LIping T, et al. "Bioresorbable polymeric stents: current status and future promise". *J. Biomater. Sci. Polym. Ed.* (2003); 14(4): 299 – 312

[65] Naguyen K, Su SH, Zilberman M, Bohluli P, Frenkel P, et al. "Biomaterials and stent technology". In *Tissue Engineering and Novel Delivery System,* ed. M Yaszemski, D Trantolo, KU Lewndrowaski, V Hasisci, D Altobelli, D Wise. (2004); 5: 107 – 30. New York: Marcel Dekker

[66] Kohn J, Langer R. "Bioresorbable and bioerodible materials". In *Biomaterials Science – An Introduction to Materials in Medicine,* ed. BD Ratner, AS Hoffman, FJ Schoen, JE Lemons, (2000); pp. 64 – 73. New Work: Academic

[67] Leenslag JW, Penning AJ, Bos RRM, Rosema FR, Boering G. "Resorbable materials of poly (L-lactic acid) VI; plates and screws for internal fracture fixation". (1987); 8: 70 – 73

[68] Bergsma EJ, Rozema FR, Bos RRM, de Bruijn WC. "Foreign body reactions to resorbable poly (L-lactide) bone plates and screws used for the fixation of unstable zygomatic fractures". *J. Oral Maxillofac. Surg.* (1993); 51: 666 – 70

[69] Bergsma EJ, de Bruijn WC, Rozema FR, Bos RRM, Boering G. "Late degradation tissue response to poly (L-lactide) bon plates and screws". *Biomaterials* (1995); 16: 25 – 31

[70] Agrawal CM, Haas KF, Leopold DA, Clark H. "Evaluation of poly (L-lactic acid) as a material for intravascular polymeric stents. *Biomaterials* (1992); 13: 176 – 82

[71] Pachence JM, Kohn J. "Bioresorbable polymers for tissue engineering". In *Principles of Tissue Engineering,* ed. RP Langer, WL Chick, (2000); pp. 267 – 70. San Diego: Academic

[72] Lewis DH. "Controlled release of bioactive agents from lactide/glycolide polymers". In *Biodegradable Polymers as Drug Delivery System,* ed. M Chasin, R Langer, (1990); pp. 1 – 41. New York: Marcel Dekker

[73] Leelarusamee N, Howard SA, Malango CJ, Ma JK. "A method for the preparation of polylactic acid microencapsules of controlled particale size and drug loading".*J. Microencapsules.*(1998); 5: 147 – 57

[74] Jeremies A, Kirtane AJ. "Balancing efficacy and safety of drug-eluting stentsin patients undergoing percutaneous coronary intervention".*Ann. Int. Med.* (2008);148:234–8

[75] Nordmann AJ, Briel M, Bucher HC. "Mortality in randomized controlled trials comparing drug-eluting vs. bare metal stents in coronary artery disease: ameta-analysis". *Euro.Heart.J.*(2006);27:2784–814

[76] Pfisterer M, Brunner-La Rocca HP, Buser PT, Rickenbacher P, HunzikerP,Mueller C, et al. "Late clinical events after Clopidogrel discontinuation may limit the benefit of drug-eluting stents: an observational study of drug-eluting versus bare-metal stents". *J. Am. Coll.Cardiol.*(2006);48:2584–91

[77] Camenzind E, Steg PG, Wijns W. "A cause for concern". *Circulation*(2007);115:1440–55

[78] Lagerqvist B, James SK, Stenestrand U, Lindbäck J, Nilsson T, Wallentin L. "Long-term outcomes with drug-eluting stents versus bare-metal stents inSweden". *N. Engl. J. Med.*(2007);356:1009–19

[79] Kastrati A, Mehilli J, Pache J, Kaiser C, Valgimigli M, Kelbäk H, et al. "Analysis of 14 trials comparing sirolimus-eluting stents with bare-metal stents". *N. Engl. J. Med.*(2007);356:1030–9

[80] Spaulding C, Daemen J, Boersma E, Cutlip DE, Serruys PW. "A pooled analysis of data comparing sirolimus-eluting stents with bare-metal stents".*N. Engl. J. Med.*(2007);356:989–97

[81]   Stettler C, Wandel S, Allemann S, KastratiA, Morice MC, Schömig A, et al. "Outcomes associated with drug-eluting and bare-metal stents: a collaborative network meta-analysis". *Lancet*(2007);370:937–48

[82]   Kirtane AJ, Gupta A, Iyengar S, Moses JW, Leon MB, Applegate R, et al. "Safety and efficacy of drug-eluting and bare metal stents: comprehensivemeta-analysis of randomized trials and observational studies". *Circulation* (2009);119:3198–206

[83]   Stone GW, Moses JW, Ellis SG, Schofer J, Dawkins KD, Morice M-C, et al. "Safety and efficacy of Sirolimus- and Paclitaxel-eluting coronary stents".*N. Engl. J. Med.* (2007);356:998–1008

[84]   Daemen J, Wenaweser P, Tsuchida K, Abrecht L, Vaina S, Morger C, et al. "Early and late coronary stent thrombosis of sirolimus-eluting and paclitaxel-eluting stents in routine clinical practice: data from a large two-institutional cohort study". *Lancet*(2007);369:667–78

[85]   Pinto T, Steinberg D, Roy P, Ashesh NB, Teruo O, Zhenyi X, et al. "Observations and outcomes of definite and probable drug-eluting stent thrombosis seen at a single hospital in a four-year period". *Am. J.Cardiol.*(2008);102:298–303

[86]   Wenaweser P, Daemen J, Zwahlen M, van Domburg R, Juni P, Vaina S, et al. "Incidence and correlates of drug-eluting stent thrombosis in routine clinical practice: 4-year results from a large 2-institutional cohort study". *J. Am. Coll. Cardiol.*(2008);52:1134–40

[87]   Park D-W, Park S-W, Park KW, Lee B-K, Kim Y-H, Lee CW, et al. "Frequency of and risk factors for stent thrombosis after drug-eluting stent implantation during long-term follow-up". *Am. J.Cardiol.*(2006);98:352–6

[88]   Iakovou I, Schmidt T, Bonizzoni E, Ge L, Sangiorgi GM, Stankovic G, et al. "Incidence, predictors, and outcome of thrombosis after successful implantation of drug-eluting stents". *JAMA*(2005);293:2126–30

[89] Finn AV, Joner M, Nakazawa G, Kolodgie F, Newell J, John MC, et al. Pathological correlates of late drug-eluting stent thrombosis: strut coverage as a marker ofendothelialization. *Circulation*(2007);115:2435–41

[90] Finn AV, Nakazawa G, Joner M, Kolodgie FD, Mont EK, Gold HK, et al. "Vascular responses to drug eluting stents: importance of delayed healing". *Arteriosc.Thromb.Vasc. Biol.*(2007);27:1500–10

[91] Joner M, Finn AV, Farb A, Mont EK, Kolodgie FD, Ladich E, et al. "Pathology of drug-eluting stents in humans: delayed healing and late thrombotic risk". *J. Am. Coll.Cardiol.*(2006);48:193–202

[92] Nakazawa G, Finn AV, Joner M, Ladich E, Kutys R, Mont EK, et al. "Delayed arterial healing and increased late stent thrombosis at culprit sites after drug-elutingstent placement for acute myocardial infarction patients: an autopsy study". *Circulation*(2008);118:1138–45

[93] Joner M, Nakazawa G, Finn AV, Quee SC, Coleman L, Acampado E, et al. "Endothelial cell recovery between comparator polymer-based drug-eluting stents". *J. Am. Coll.Cardiol.*(2008);52:333–42

[94] Talija M, Valimaa T, Tammela T, Petas A, Tormala P. "Bioabsorable and biodegradable stents in urology". *J. Endourol.* (1997); 11: 391 – 97

[95] Kapoor R, Liatsikos EN, Badlani G. "Endoprostatic stents for management of benign prostatic hyperplasia". *Curr.Opin. Urol.* (2000); 10: 19 – 22

[96] Kletscher BA, Oesterling JE. "Prostatic stents: current perspective for the management of benign prostatic hyperplasia". *Urol. Clin. North Am.* (1995); 22: 423 – 30

[97] de la Rosette JJMC, Beerlage HP, Debruyne FMJ. "Role of temporary stents in alternative treatment of benign prostatic hyperplasia".*J. Endourol.* (1997); 11: 467 – 72

[98] Lumiaho J, Heino A, Pietilainen T, Ala-Opas M, Talja M, et al. " The morphological, in situ effect of a self-reinforced bioabsorbablepolylactide (SR – PLA 96) ureteric stent; an experimental study." *J. Urol.* (2000); 164: 1360 – 63

[99] Lumiaho J, Heino A, Tunninen V, Ala-Opas M, Talja M, et al. "New bioabsorbablepolylactide ureteral stent in the treatment of ureteral lesions: an experimental study. *J. Endourol.* (1999); 13: 107 – 12

[100] Rafana AL, Mehta AC. "Stenting of the tracheobronchial tree".*Radiol.Clin. North Am.* (2000); 38: 395 – 408

[101] Robey TC, Eiselt PM, Murphy HS, Mooney DJ, Weatherly RA."Biodegradable external tracheal stents and their use in a rabbit tracheal reconstruction model".*Laryngoscope* (2000); 110: 1936 – 42

[102] Lochbihler H, Hoelzi J, Dietz H. "Tissue compatibility and biodegradation of new absorbable stents for tracheal stabilization: an experimental study". *J. Pediatr. Surg.* (1997); 32: 717 – 20

[103] Filler RM, Forte V, Farga JC, Matute J. "The use of expandable metallic airway stents for tracheobronchial obstruction in children".*J. Pediatr. Surg.* (1995); 30: 1050 – 55

[104] Furman RH, Backer CL, Dunham M, Donaldson J, Mavroudis C, Holinger LD. "The use of balloon-expandable metallic stents in the treatment of pediatric tracheomalacia and bronchomalacia".*Arcb.Otolaryngol. Head neck Surg.* (1999); 125: 203 – 7

[105] Lamber R. "Treatment of esophagogastric tumors".*Endoscopy* (2000); 32: 322 – 31

[106] Morgan R, Adam A. "Use of metallic stents and balloons in the esophagus and gastrointestinal tract".*J. Vasc.Interv.Radiol.* (2001); 12: 283 – 97

[107] Nicholson T. "Other uses of nonvascular stents".*Hosp. Med.* (2000); 61: 97 – 102

[108] Prabhu S, Hossainy S. "Modeling of degradation and drug release from a biodegradable stent coating".*J. Biomed. Mat. Research A* (2006); doi: 10.1002/jbm.a.31053

[109] Prabhu S, Hossainy S. "A mathematical modeling for predicting drug release from a biodurable drug-eluting stent coating".*J. Biomed. Mat. Research A* (2008); doi: 10.1002/jbm.a.31787

[110] Hugh QZ, Dudley J, Hossainy S, Lewis BS. "A theoretical model to characterize the drug release behavior of drug-eluting stents with durable polymer matrix coating". *J. Biomed. Mat. Research A* (2011); doi: 10.1002/jbm.a.33246

[111] Versteeg HK, Malalasekera W. "An Introduction to Computational Fluid Dynamics: The Finite Volume Method" in Longman Scientific &Technical & (1995), chapter 4, 7 & 8. pp. 86 – 87, 99 – 100, 159, 169 – 172, Jhon Willy & sons Inc. 605 Third Avenue, New Work, NY 10158

Appendix A

SOME NECESSARY THEORIES OF FINITE VOLUME METHOD

Computational fluid dynamics or CFD is the analysis of system involving fluid flow, heat transfer and associated phenomena like mass transfer and chemical reaction; by means of computer based simulation. CFD uses powerful computers and applied mathematics to model fluid flow situations. It has three major application field, Industrial, environmental and physiological application. In case of industrial application CFD works with aerospace, automotive, biomedical, chemical & process, combustion, electronic & computer, marine, mechanical, metallurgical, nuclear engineering, architecture, glass manufacturing, petroleum engineering, power generation, train & turbo machinery design, water engineering etc. In case of environmental application it works with atmosphere pollution, climate calculations, fire in buildings, oceanic flows, pollution of natural water and safety. In the field of physiology CFD can deal with cardiovascular flow and flow in lungs and breathing passages. The main advantage of using CFD can be remarked as follows:

- It provides detailed understanding of flow distribution, weight loss, mass and heat transfer, particulate separation, etc. Consequently, all these give plant managers a much better and deeper understanding of what is happening in a particular process or system.
- It can make possible evaluation of geometric changes with much less time and cost than laboratory working.
- It can answer many 'what if' questions in short time by simulation of models with changed parameters,
- It is able to reduce scale-up problems because the model is based on fundamental physics and are scale independent.
- It is particularly useful in simulating conditions where it is not possible to take detailed measurements such as high temperature or dangerous environment in an oven.
- Since it is a pro-active analysis and design tool, it can highlight the root cause not just the effect when evaluating plant problems.

A.1 Finite Volume Method for One-dimensional Unsteady State Diffusion

The conservation law for the transport of a scalar Ø, in an unsteady flow has the following general form

$$\frac{\partial}{\partial t}(\rho\emptyset) + div(\rho u\emptyset) = div(\Gamma\, grad\phi) + S_\phi$$

The first term of the equation represents the rate of change term of Ø which is zero for steady flow. Second term represent the change in Ø due to convection and third term shoes diffusion of Ø where fourth term give source of generation or consumption sink of Ø. The integration based numerical solution of this equation within a control volume is mainly the finite volume method. The control volume integration, which forms the key step of the finite volume method, distinguishes it from all other CFD techniques, yielding following form.

$$\int_{CV} \frac{\partial}{\partial t}(\rho\emptyset) + \int_{CV} div(\rho u\emptyset) = \int_{CV} div(\Gamma\, grad\phi) + \int_{CV} S_\phi$$

By working with one dimensional unsteady state diffusion equation, the approximation techniques that are needed to obtain the discretized equation is as follows.

The unsteady diffusion of a property Ø in one-dimensional domain can be defined by figure A.1. The process is governed by,

$$\rho\frac{\partial\emptyset}{\partial t} = \frac{\partial}{\partial x}\left(\Gamma\frac{\partial\emptyset}{\partial x}\right) + S \qquad (A.1)$$

Where, Γ is the diffusion coefficient. Boundary values of Ø at point A and B are prescribed.

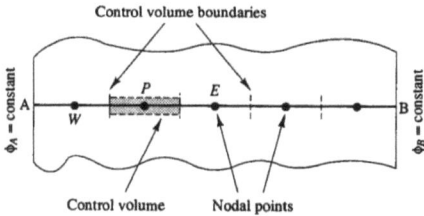

Fig. A.1: One-dimensional domain of control volume

A.1.1 Grid Generation of the Finite Volume

The first step in the finite volume method is to divide the domain into discrete control volumes. By putting node points between A and B as shown in figure A.2, boundaries of the control volumes are positioned mid-way midway between adjacent node. Thus each node is surrounded by a control volume or cell. It is common practice to setup control volumes near the edge of the domain in such a way that physical boundaries coincide with the control volume boundaries. The usual convention of node point notation in CFD has been shown in figure A.2.

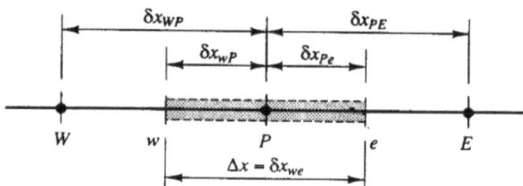

Fig. A.2: Control volume node point in one-dimensional domain

A general node is identified with P and its neighbor in one dimensional geometry; the nodes to west and east are identified by W and E respectively. The west side face of the control volume is referred to by 'w' and the east side control volume face by 'e'. The distance between node W and P and between nodes P and E, are identified by Δx_{WP} and Δx_{PE} respectively. Similarly the distance between face w and node P and face e and node P is identified as Δx_{wP} and Δx_{eP} respectively, where $\Delta x_{wP} = \Delta x_{eP} = \frac{\Delta x_{WP}}{2} = \frac{\Delta x_{PE}}{2}$ and $\Delta x_{WP} = \Delta x_{PE} = \Delta x_{we} = \Delta x$.

A.1.2 Discretization of the Integral Equation

The key step of the finite volume method is the integration of the governing equation over a control volume to yield a discretized equation at its nodal point P. For the control volume defined above the discretization will be as follows of unsteady 1D diffusion. Integration of equation A.1 over the control volume shown in figure A.2 and over a time interval from t to t + Δt gives

$$\int_{t}^{t+\Delta t}\int_{CV} \rho \frac{\partial \emptyset}{\partial t} dV\, dt = \int_{t}^{t+\Delta t}\int_{CV} \frac{\partial}{\partial x}\left(\Gamma \frac{\partial \emptyset}{\partial x}\right) dV\, dt + \int_{t}^{t+\Delta t}\int_{CV} s\, dV\, dt$$

This may be written as

$$\int_w^e \left[\int_t^{t+\Delta t} \rho \frac{\partial \emptyset}{\partial t} dt\right] dV = \int_t^{t+\Delta t} \left[\left(\Gamma A \frac{\partial \emptyset}{\partial x}\right)_e - \left(\Gamma A \frac{\partial \emptyset}{\partial x}\right)_w\right] dt + \int_t^{t+\Delta t} S \Delta V \, dt \quad (A.2)$$

In this equation A is the face area of the control volume, ΔV is its volume, where, $\Delta V = A \Delta x$ and Δx is the width of the control volume and S is the average source strength. If $\emptyset$ at a node is assumed to prevail over the whole control volume, the left side of equation A.2 can be written as

$$\int_w^e \left[\int_t^{t+\Delta t} \rho \frac{\partial \emptyset}{\partial t} dt\right] dV = \rho(\emptyset_P - \emptyset_P^0) \Delta V$$

Here, superscript 0 refers to the $\emptyset$ at time t; $\emptyset$ at time level $t + \Delta t$ are not superscripted. Thus equation A.2 become,

$$\rho(\emptyset_P - \emptyset_P^0)\Delta V = \int_t^{t+\Delta t} \left[\left(\Gamma_e A \frac{\emptyset_E - \emptyset_P}{\Delta x_{PE}}\right) - \left(\Gamma_w A \frac{\emptyset_P - \emptyset_W}{\Delta x_{WP}}\right)\right] dt + \int_t^{t+\Delta t} S \Delta V \, dt \quad (A.3)$$

The time integrals at the right side of equation A.3 is evaluated by the following way,

$$I_T = \int_t^{t+\Delta t} \emptyset_P dt = [\theta \emptyset_P + (1-\theta)\emptyset_P^0]\Delta t$$

Thus equation A.3 become,

$$\rho\left(\frac{\emptyset_P - \emptyset_P^0}{\Delta t}\right)\Delta x$$
$$= \theta\left[\left(\Gamma_e \frac{\emptyset_E - \emptyset_P}{\Delta x_{PE}}\right) - \left(\Gamma_w \frac{\emptyset_P - \emptyset_W}{\Delta x_{WP}}\right)\right]$$
$$+ (1-\theta)\left[\left(\Gamma_e A \frac{\emptyset_E^0 - \emptyset_P^0}{\Delta x_{PE}}\right) - \left(\Gamma_w A \frac{\emptyset_P^0 - \emptyset_W^0}{\Delta x_{WP}}\right)\right] + S\Delta x$$

This equation can be rearranged as follows

$$\left[\rho\frac{\Delta x}{\Delta t}+\theta\left(\frac{\Gamma_e}{\Delta x_{PE}}+\frac{\Gamma_w}{\Delta x_{WP}}\right)\right]\emptyset_P$$

$$=\frac{\Gamma_e}{\Delta x_{PE}}[\theta\emptyset_E+(1-\theta)\emptyset_E^0]+\frac{\Gamma_w}{\Delta x_{WP}}[\theta\emptyset_W+(1-\theta)\emptyset_W^0]$$

$$+\left[\rho\frac{\Delta x}{\Delta t}-(1-\theta)\left(\frac{\Gamma_e}{\Delta x_{PE}}+\frac{\Gamma_w}{\Delta x_{WP}}\right)\right]\emptyset_P^0+S\Delta x$$

This equation can be written as

$$a_P\emptyset_P = a_E[\theta\emptyset_E+(1-\theta)\emptyset_E^0]+a_W[\theta\emptyset_W+(1-\theta)\emptyset_W^0]$$
$$+[a_P^0-(1-\theta)(a_E+a_W)]\emptyset_P^0+b \qquad (A.4)$$

where, $a_E = \frac{\Gamma_e}{\Delta x_{PE}}$, $a_W = \frac{\Gamma_w}{\Delta x_{WP}}$, $b = S\Delta x$, $a_P^0 = \rho\frac{\Delta x}{\Delta t}$ and $a_P = a_P^0 + \theta(a_E+a_W)$

A.1.2.1 Explicit Scheme

In explicit scheme the source term is linearized as b = S_W + $S_P\emptyset_P^0$ and θ is set to zero. Thus equation A.4 become,

$$a_P\emptyset_P = a_E\emptyset_E^0+a_W\emptyset_W^0+[a_P^0-(a_E+a_W-S_P)]\emptyset_P^0+S_u \qquad (A.5)$$

Where, $a_P = a_P^0$

Right hand side of the equation contain only the Ø terms of previous time step, left hand side can be calculated by forward marching in time. The scheme is based on backward differencing and its Taylor series truncation error accuracy in first order with time. The coefficient of $\emptyset_P^0$ may be viewed as neighbor coefficient connecting the values at the old time level to those at the new time level. For this coefficient to be positive $a_P^0 - a_E - a_W$ should be greater than zero. Thus for uniform grid spacing, limiting condition for the explicit scheme will be,

$$\rho\frac{\Delta x}{\Delta t} > \frac{2\Gamma}{\Delta x} \Rightarrow \Delta t < \rho\frac{(\Delta x)^2}{2\Gamma}$$

The above inequality set a stringent maximum limit to the time step size and represents a serious limitation for explicit scheme.

A.1.2.2 Crank-Nicolson Scheme

Crank-Nicolson scheme results from setting $\theta = \frac{1}{2}$ in equation A.4. Thus the discretized equation become,

$$a_P \emptyset_P = a_E \left[\frac{\emptyset_E + \emptyset_E^0}{2}\right] + a_W \left[\frac{\emptyset_W + \emptyset_W^0}{2}\right] + \left[a_P^0 - \frac{a_E}{2} - \frac{a_W}{2}\right] \emptyset_P^0 + b \qquad (A.6)$$

Where,

$$a_P = \frac{1}{2}(a_E + a_W) + a_P^0 - \frac{1}{2}S_P \quad and \quad b = S_u + \frac{1}{2}S_P \emptyset_P^0$$

As more than one $\emptyset$ term is present in equation A.6 thus simultaneous equations at all node point have to be solved at each time step. To ensure all positive coefficients in equation A.6, condition is $a_P^0 > \left[\frac{a_E}{2} + \frac{a_W}{2}\right]$. This leads to

$$\Delta t < \rho \frac{(\Delta x)^2}{\Gamma}$$

This time step limitation is slightly less restrictive than the limitation of explicit scheme. This method is based on central differencing scheme and hence is second order time accurate.

A.1.2.3 Fully Implicit Scheme

When the value of θ in equation A.4 is set to 1 then fully implicit scheme is obtained. The discretized equation here is,

$$a_P \emptyset_P = a_E \emptyset_E + a_W \emptyset_W + a_P^0 \emptyset_P^0 + S_u \qquad (A.7)$$

Where, $a_P = a_E + a_W + a_P^0 - S_P$

As more than $\emptyset$ terms present in equation A.6 is at present time step except $\emptyset_P^0$ thus simultaneous equations at all node point have to be solved at each time step. Again all coefficients present in this equation are positive. Thus there is no time step or space step limitation in fully implicit method.

Discretized equation of any algorithm is than solved using TDMA.

A.2 Finite Volume Method for Two-dimensional Unsteady State Diffusion

The unsteady diffusion of a property Ø in one-dimensional domain can be defined by figure A.3.

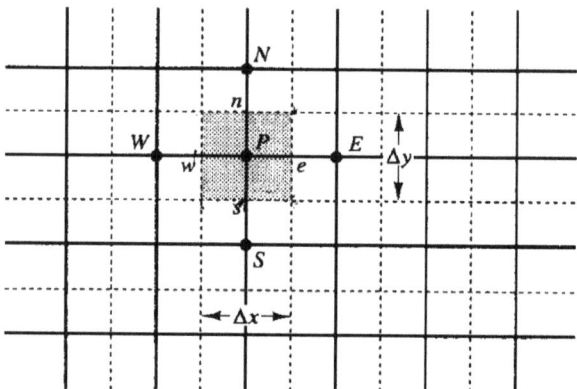

Fig. A.3: Part of two-dimensional grid

The process is governed by,

$$\rho \frac{\partial \emptyset}{\partial t} = \frac{\partial}{\partial x}\left(\Gamma \frac{\partial \emptyset}{\partial x}\right) + \frac{\partial}{\partial y}\left(\Gamma \frac{\partial \emptyset}{\partial y}\right) + S \qquad (A.8)$$

From which final form of discretized equation is found to be, with fully implicit method consideration,

$$a_P \emptyset_P = a_E \emptyset_E + a_W \emptyset_W + a_N \emptyset_N + a_S \emptyset_S + a_P^0 \emptyset_P^0 + b$$

To solve above discretized equation using TDMA, the equation is rearranged by the following way,

$$-a_W \emptyset_W + a_P \emptyset_P - a_E \emptyset_E = a_N \emptyset_N + a_S \emptyset_S + a_P^0 \emptyset_P^0 + b$$

The right hand side of this equation is assumed to be temporarily known and then solved along n – s direction for the chosen line shown in figure A.4

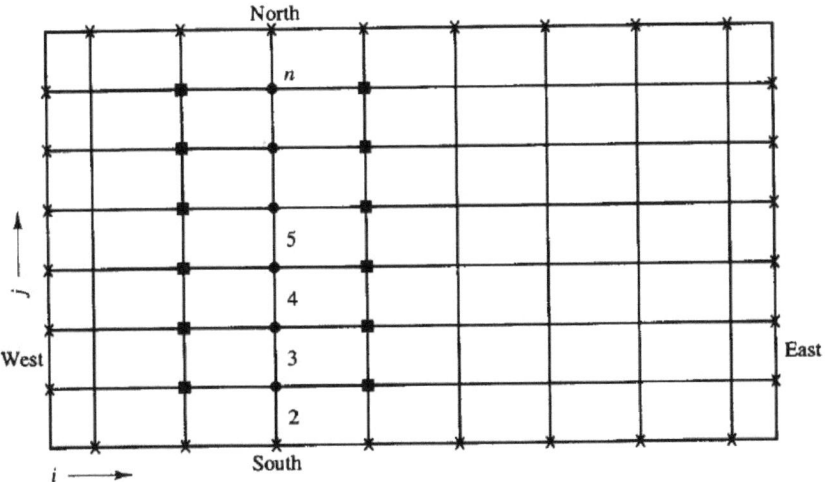

- Points at which values are calculated
- Points at which values are considered to be temporarily known
- x Known boundary values

Fig. A.4: Line by line application of TDMA method.

Subsequently the calculation is moved to the next north-south line. The sequence in which lines are chosen is known as the sweep direction. If sweep is done from west to east the value of $Ø_W$ to the west of point P are known from the calculation on the previous line. The value of $Ø_E$ as unknown thus the solution must have to be iterative. At each iteration step the value of $Ø_E$ is taken from the value of $Ø_E$ at the previous calculation step. This iteration is repeated until a converged solution is obtained.

ABOUT THE AUTHORS

Md. Iqbal Hossain
Editor

Md. Iqbal Hossain obtained Ph.D. from the School of Chemical and Biomedical Engineering, Nanyang Technological University, Singapore and has been serving as an Assistant Professor at the Department of Chemical Engineering, Bangladesh University of Engineering and Technology. He is also a Senior Member of the American Institute of Chemical Engineers. His areas of expertise are multiphase flow, hydrodynamics and measurement-computational techniques.
Email: iqbalhossain@che.buet.ac.bd

Nadia Sultana
Author

Nadia Sultana obtained B.Sc. and M.Sc. from the Department of Chemical Engineering, Bangladesh University of Engineering and Technology. She obtained her second M.Sc. from the Department of Chemical Engineering, Texas Tech University. She is currently a Ph.D. Candidate at Texas Tech University. In her first M.Sc. degree, her research focused on the simulation of a 2D mass transfer profile in a CAD drug eluting stent. In her second M.Sc. degree, she worked

with a 3D alginate gel micro-bead (LbL surface modified) based microenvironment for 3T3-L1 cell cultures. Currently, she is researching on the process of heterogeneous crystal nucleation for the energetic material 2, 4, 6-trinitrotoluene (TNT).